D0312231

How to Be Human in the Digital Economy

How to Be Human in the Digital Economy

Nicholas Agar

The MIT Press
Cambridge, Massachusetts
London, England

© 2019 Massachusetts Institute of Technology

All rights reserved. No part of this book may be reproduced in any form by any electronic or mechanical means (including photocopying, recording, or information storage and retrieval) without permission in writing from the publisher.

This book was set in Stone Serif by Westchester Publishing Services. Printed and bound in the United States of America.

Library of Congress Cataloging-in-Publication Data

Names: Agar, Nicholas, author.
Title: How to be human in the digital economy / Nicholas Agar.
Description: Cambridge, MA : MIT Press, 2018. | Includes bibliographical references and index.
Identifiers: LCCN 2018007542 | ISBN 9780262038744 (hardcover : alk. paper)
Subjects: LCSH: Internet--Social aspects. | Information technology--Social aspects. | Agent (Philosophy)
Classification: LCC HM851 .A333 2018 | DDC 302.23/1--dc23 LC record available at https://lccn.loc.gov/2018007542

10 9 8 7 6 5 4 3 2 1

Contents

Acknowledgments ix

Introduction: Taking the Long View of Digital Revolution 1

The Threat to Human Agency 3

We Should Avoid a Present Bias about Computers and a Belief in Human Exceptionalism 6

Forward to a Social-Digital Future 9

A Note on Philosophical Method 15

An Outline of the Book 16

1 Is the Digital Revolution the Next Big Thing? 23

Will the Digital Revolution Fizzle? 26

The Magic Combination of Artificial Intelligence and Data 30

How AI Could Transform Transportation 33

How AI Could Transform Health 38

Concluding Comments 42

2 AI's Split Personality—Minds or Mind Workers? 43

Philosophical and Pragmatic Interests in Machine Minds: A Focus on Making Minds or on Doing Mind Work 45

The Difference between Authentic and Ersatz Minds 51

Hyperactive Agency Detectors and Human-Like Machines 54

A Moral Reason to Avoid Creating Machines with Minds 57

Concluding Comments 59

3 Data as a New Form of Wealth 61

How Could Data Be Wealth? 62

Unfairness and the New Forms of Wealth 66

Does Data Want to Be Free? 68

Do unto Facebook and Google … Micropayments for the Use of Our Data? 73

Concluding Comments 79

4 **Can Work Be a Norm for Humans in the Digital Age?** 81
Searching for Work that Is Both Productive and Therapeutic in the Digital Age 82
The Inductive Optimism of the Economists 84
The Protean Powers of the Digital Package 88
Will Humans Always Control the Last Mile of Choice? 92
A Conjecture about the Labor Market of the Digital Age 99
Gaining Philosophical Perspective on the Dispute between Optimists and Pessimists 102
Concluding Comments 106

5 **Caring about the Feelings of Lovers and Baristas** 109
What Is It Like to Love a Robot? 110
From Romantic to Work Relationships 118
What Counts as a Social Job? 123
Can I Justify My Pro-Human Bias? 125
Concluding Comments 129

6 **Features of the Social Economy in the Digital Age** 131
Two Economies for the Digital Age 132
Some Noteworthy Differences between Social and Digital Goods 135
The Ambiguous Digital Futures of Sales Assistants 143
The Different Digital Age Futures of Uber and Airbnb 145
Space Exploration as Social Work 149
Concluding Comments 153

7 **A Tempered Optimism about the Digital Age** 155
The Different Logic of Predictions and Ideals 156
We Should Prefer Robust Ideals 162
The Social-Digital Economy versus the Collaborative Commons 163
The Social-Digital Economy versus a Jobless Future with a Universal Basic Income 165
The UBI as an Inadequate Response to Inequality in the Digital Age 167
An Expanded Basic Income? 171
Concluding Comments 173

8 **Machine Breaking for the Digital Age** 175
See through the Digital Halo Effect! 176
Don't Fall for Tech TINA! 179
If You Can Cheat an Algorithm, Then Why Not? 181
Work for Free for Oxfam, but Make Facebook Pay! 183
Don't Fight the Last War! 184
Concluding Comments 189

9 Making a Very Human Digital Age 191
Welcoming a Social Age 196

Notes 199
Index 217

Acknowledgments

I received invaluable suggestions and support from many people in the writing of this book. Colleagues, students, and friends read the manuscript or commented incisively on the book's central ideas. I thank Daniel Agar, Jan Agar, Fin Ashworth, Pablo Barranquero, Billie Berry, Stuart Brock, Lucy Campbell, Phil Cook, Jonette Crysell, David Dwyer, Anthony Hall, David Lawrence, Simon Keller, Tom Malone, Edwin Mares, Cei Maslen, Johnny McDonald, Jonathan Pengelly, Sandra Park, Yiyan Wang, and Jennifer Windsor.

Anonymous readers for the Press offered comments leading to many improvements. I hope they can see how their suggestions made the book's central claims more persuasive.

I'm grateful to Phil Laughlin who expertly shepherded the book through from emailed proposal to printed book. I'm also thankful to Marcy Ross and Bridget Leahy for correcting many awkward sentences.

I owe a different kind of gratitude to the staff at the Aro Street Café, the Brooklyn Deli, and The Bresolin, who kept me supplied with coffee during the writing of these chapters and knew to cut me off at the earliest sign of caffeine psychosis.

Finally, I thank my fantastic wife Laurianne and my magnificent kids Alexei and Rafael. I hope Alexei and Rafael get to enjoy the fruits of a genuinely social age.

Introduction: Taking the Long View of Digital Revolution

The Digital Revolution is transforming human lives. Here I define the Digital Revolution as the widespread and rapid replacement of mechanical and analog electronic technologies by digital technologies. Ground zero for these upheavals and transformations is the digital computer. But the revolution's effects reach far beyond the stereotypical desktop word processor. Digital technologies are radically changing the ways that we share information, travel, treat disease, and party. An observable acceleration in the power of computers suggests that the transformations and dislocations we are currently experiencing are just the beginning.

Technological revolutions are, in the account presented in this book, more than just interesting events in human history. They are history's motors. The Renaissance jurist, statesman, and advocate of science, Francis Bacon, made an emphatic statement about the significance of technological advances to human affairs. Writing about the significance of his own age's big technological innovations—"printing, gunpowder, and the magnet"—he said "no empire, no sect, no star seems to have exerted greater power and influence in human affairs than these mechanical discoveries."[1] The Digital Revolution is influencing human affairs and seems set to have even greater influence. Thrilling gadgets and apps are merely its most visible manifestations.

We have a choice about how to view the Digital Revolution. This choice arises from the fact that it is a complex, multi-strand event that comprises many individuals and groups of individuals, and many technologies and categories of technology. The Digital Revolution's origin is difficult to determine with any precision. We know that its early stages featured the theoretical insights of Alan Turing and the tinkering of various geniuses at Bell

Labs and Xerox Park in the middle years of the twentieth century.[2] Perhaps its beginning can be traced back as far as Charles Babbage and Ada Lovelace in nineteenth-century England. Babbage was the first person to attempt (unsuccessfully) to build something that we today would acknowledge as a computer and Lovelace may have been the inventor of computer programming. A great deal has happened since these modest beginnings.

This book takes the long view of the Digital Revolution. The long view permits us to address questions about where digital technologies are taking us. It is not overly concerned about the specifics of today's buzz-worthy digital technologies. In the long view, the Macintosh computer, the Twitter social networking platform, and the Oculus Rift virtual reality headset feature only as undifferentiated parts of the onrushing blur of progress in digital technologies. An enhanced sense of the broad meaning of technological change compensates for the long view's lack of detail. It offers an awareness of the grand sweep of technological change and its implications for human affairs. No fact about how integrated circuits process information describes their lasting effects on human experience and on the arrangements of the societies that host them. In the long view the Digital Revolution is the latest event in a historical sequence that begins with the Neolithic Revolution—which brought about farming, permanent settlements, and social hierarchies—and progresses to the Industrial Revolution, which brought on mechanization, mass production, and globalization. When we take the long view, we hope to see what is truly lasting about the Digital Revolution. Once we have gotten used to, or recovered from, the serial shocks of all these digital novelties, what effects will the Digital Revolution have on humanity?

The focus of the long view on trends means that it has little to say about some of our more immediate concerns. Suppose that you hear someone shouting "Help, I'm being assaulted!" It's not particularly useful to respond "Don't worry! Crime in this area is down 80 percent." Information about crime trends does nothing to address the immediate concerns of someone who is currently being mugged. Similarly, pointing to long-term trends does not offer much to someone who has just been put out of work by automation. The long view nevertheless offers something missing from accounts that focus on immediate effects. It directs attention away from the digital trees to gain a better understanding of the digital forest. A metaphorical squinting of the eyes brings more general truths into focus.

In pictures of Earth taken from space we can see the largest human-made objects—the Great Wall of China, the Pyramids of Giza, and Dubai's Palm Island. Also apparent are some of the patterns of human activity—the intense night-time lights of the cities of North America and Europe and the eerie absence of lighting in North Korea's night sky. At this scale, however, individual humans are invisible. Something similar happens in the long view of technological change. We lose sight of individuals. A history of the Industrial Revolution may tell us about the 204 miners who died in the Hartley Colliery disaster in 1862 when they were trapped below after an accident with the pit's pump. We may have some abstract appreciation of the suffering caused to them and their families. But for people of the early twenty-first century interested in the long view of technological change, that disaster is mainly interesting for the lasting reforms it prompted. It led to legislation requiring coal mines to have at least two independent paths of escape. Viewed from the second decade of the twenty-first century, the miners of Hartley Colliery are dead and gone whether the disaster occurred or not.

The Threat to Human Agency

If we are interested in what is truly lasting and unprecedented about the Digital Revolution, we must look beyond the consternation and excitement that are predictable consequences of technological churn. We recover from some technological shocks promptly and fully. The Harrods department store offered brandy and smelling salts to help people to recover from the shock of a ride on its first escalators, installed in 1898. Some of today's escalator travelers might enjoy a brandy upon arrival but it's not something we need to recover from a ride to a department store's upper floor. My principal focus is on a challenge to human agency. I argue that the Digital Revolution poses a threat to humans as doers, as authors of our own destinies. We make significant choices about ourselves and the world, in large part because we reason in certain ways.

There are two ways to think about this challenge to agency. First, there is a threat to human agency writ small. Here the economic value of human agency is the principal target. Can humans keep their jobs if forced to compete against machines capable of performing every job-related task better and more cheaply? Fears about technological unemployment have led to forecasts of the abolition of a disparate collection of jobs.[3] On one such list, waiters and waitresses face a 90 percent risk of finding their work automated by 2035.[4] Technological unemployment resulting from the Digital Revolution is no respecter of social station. The threat for chartered and certified accountants over the same period is still greater at 95 percent. Twenty years may be long enough for many to reach retirement age, but they should think twice about urging daughters or sons to follow them into the family trade. Pointing to a favored few professions that seem to be immune to replacement by digital technologies does not respond to this threat. Perhaps humans will remain supreme in the production of abstract art or the performance of stand-up comedy. But many human workers seem both eminently and imminently replaceable by digital machines.

Much of the threat to human agency emerges from developments in artificial intelligence and more specifically in machine learning, a field that aims at producing machines capable of learning without explicit instruction from human programmers. We see precursors of a future in which human agency has lost much of its value in the autopilots that take an increasing responsibility for flying our passenger jets, the driverless cars that are likely to be soon traversing our highways more efficiently and safely than cars driven by humans, and the computers that we entrust with finding patterns in data about who does and who doesn't get melanoma that would be beyond the perception of the most dedicated and perceptive human medical researchers. Much of the disruption caused by the Industrial Revolution came from its automation of muscle power. A power loom run by a comparatively unskilled operator did the work of many skilled handloom weavers. The Digital Revolution is automating human mind work. Perhaps we will conclude that the computers that displace human mind workers are themselves mindless. But even so, they mindlessly do human mind work. The future promises digital machines that do this work both to a higher standard and more cheaply.

Mind work is not an all-or-nothing category. All work that humans do requires the use of our minds. A worker whose sole task is stacking bricks would be completely unable to do her job if she could not understand the instructions given her about where to stack them. But there are some jobs whose intellectual content is higher than others. Mental labor makes a smaller contribution to brick-stacking than it does to accounting or investigative journalism. The Digital Revolution poses a threat to jobs whose intellectual content is high, jobs which typically demand prolonged educations and provide high rates of pay.

Perhaps there will always be some things that humans can do that computers cannot. Humans can not only work but also whistle while they work, whereas computers, famed for their multitasking, may never dual task in precisely this way. The real threat to the human job applicants of the Digital Age concerns the economic value of that whistling. We are talking about potential futures in which employers choose to forgo the whistling to get more work done faster and more cheaply.

Second, there is the threat to human agency writ large. What's at issue here is our control over our collective destiny. The civilization that emerged from the Industrial Revolution was, like the one that preceded it, one in which humans made the key choices. Technological change redirected human agency but did not make it irrelevant. Advances in artificial intelligence seem to lead to a progressive erosion of human agency. We seem to face a future in which control over human societies and lives is incrementally and inexorably surrendered to digital technologies with manifestly superior powers of decision-making. The decisions we make about how to get to an unfamiliar destination will be restricted to the decision to speak its name into the navigation system of a driverless car. Perhaps the automated deciders of the Digital Age will respond to today's violent and competing populisms by solving our big problems for us. Will we address the problem of climate change by complying with the commands of a machine intelligence with access to the totality of data about the global climate system?

Consider some thoughts of Apple co-founder Steve Wozniak. Wozniak was responding to a fear, expressed by the astrophysicist Stephen Hawking and the philosopher Nick Bostrom, that advances in artificial intelligence might turn the plot line of the *Terminator* movies into reality.[5] We would create an AI with the capacity to improve itself. This AI would rapidly become more powerful than all humanity combined. It might decide that the world is a better place without us.[6] Wozniak took a rosier attitude toward these imminent digital superintelligences. Speaking in 2015, Wozniak said of these expected future machine intelligences: "They're going to be smarter than us and if they're smarter than us then they'll realize they need us." In Wozniak's vision of the future, humans are not incinerated by AI-triggered nuclear exchanges. Rather we become cherished and mollycoddled pets of superintelligent AIs. "We want to be the family pet and be taken care of all the time."[7] The artificial superintelligences that will predictably emerge from the Digital Revolution will cater to this need. Wozniak reflects on the "filet steak and chicken" he feeds his dog, clearly relishing the delights that a future AI might cook up for him.

Wozniak's vision of our future relationship with artificial superintelligences is sunnier than the *Terminator* vision. Much better for the humans of the Digital Age to be like dogs dining on filet steak than those deemed more trouble than they are worth and administered lethal doses of pentobarbital. However, the Wozniak and Hawking visions are equal affronts to those who hope for a vision of the future in which humans retain authority over the machines and over our own destinies.[8] One 2012 estimate placed the global number of dogs at 525 million.[9] The countless choices made by these animals have zero consequences for the destiny of global civilization. It seems better to play the role of pampered poodle than to be incinerated in a AI-triggered nuclear exchange, but both are visions of the future in which we surrender authority over our collective destinies. In times of high stress, you may find yourself looking at your contentedly dozing pet and uttering the words "I wish I was a dog!" But in times of greater confidence we fight for our control over our destinies, both as individuals, and collectively.

These forecasts of a radically disempowered humanity may seem the stuff of science fiction. Yet they are predictable consequences of the development of digital technologies. The long view of the Digital Revolution directs attention away from today's limited digital deciders. It focuses on what they will predictably become. It warns against a dehumanized future dominated by the value of efficiency in which we realize that we do better without other humans. It's effectively an indolent path to extinction in which we incrementally cede our places to our robotic betters.

We Should Avoid a Present Bias about Computers and a Belief in Human Exceptionalism

If we are to come to terms with the threat to human agency from computers, then we must avoid a bias in our assessment of future threats. Humans have a built-in tendency to suppose that things will continue as they are now. We tend to be overly influenced by the evidence presented directly to our senses. The experts tell us that we face a civilizational threat from climate change. But our beach houses are not yet under water and the shelves of our supermarkets are laden with fresh vegetables. The climate is supposed to be warming. But this morning was very chilly indeed. Even those who intellectually accept the danger of climate change fail to muster the kind of response the problem demands. We find it difficult to accept that the future could be so bad when everything now seems just fine.

This *present bias* leads us to understate the machines' threat to human agency. Many of today's artificial intelligences are quite stupid. These are no threat to our jobs. When we take the long view, we compare what humans may become with what machines may become. Humans can improve—our flexible brains permit us to learn new tricks. But the rate of improvement of the machines is especially steep. The failings of present machines should not blind us to the expected capacities of their imminent descendants. Our position in respect of the machines that may take our jobs may be analogous to chess grandmasters in the early 1990s. We should combine arrogance about the capacities of some of today's computers with humility in the face of the capacities of their descendants.

As we seek to automate familiar human tasks, we are especially aware of the machines' mishaps. But human drivers are also far from perfect; we are accustomed to their mishaps. The programmed oversights of self-driving cars make world news.[10] Unless they kill a princess, the lethal errors of human drivers don't attract worldwide attention. No digital technology is logically immune to error. But perfection is not a standard we should expect. Perfect safety is unachievable, but the standard of being much safer than human drivers is both achievable and expected. Perhaps some of Deep Blue's programmers fantasized about a machine that could play perfect chess, a machine that could demand the resignation of its opponent before its first move by presenting a detailed description of her inevitable checkmate. This may not be possible. But, as Deep Blue and its successors have shown, the standard of playing better chess than the best human player certainly is. The machines can't beat God, the hypothesized perfect chess player, but they can beat any human.

The present bias against the capacities of future machines combines with an exaggeration of our own abilities. I call this bias a *belief in human exceptionalism*. A sober comparison of today's humans with today's computers reveals things that we do with ease that computers are hopeless at. Believers in human exceptionalism accept that computers have already overtaken us in many areas and are fast gaining on us in others. But they insist on a core of human capacities that will remain forever out of the machines' reach. These will keep us employed in an age of ubiquitous supercomputers.

A belief in human exceptionalism inclines us to prefer quasi mystical names for our most cherished mental capacities. Machine thinkers comply with algorithms. Human thinkers possess genius, they intuit answers, and they demonstrate wisdom. "Genius," "intuit," and "wisdom" may be acceptable as brand names for digital products. But we resist the suggestions that computers could manifest these traits for real. If we accept this reasoning, a computer may execute a billion computations per second, but it can never be wise. Popular culture panders to this belief in human exceptionalism. Captain Kirk outwits an ostensibly superior alien intellect by proposing to implement the "Corbomite Maneuver," a piece of nonsense invented by his human brain that confuses the alien. Kirk tells the aggressor that the *Enterprise* contains corbomite—an unexplained substance that destroys any attacker. The alien's logic-limited brain seems to prevent it from seeing through this ruse.

The belief in human exceptionalism purports to construct a barrier that would prevent machines from replicating or mimicking one of the celebrated achievements of the human scientific imagination. The German chemist August Kekulé was trying to figure out the structure of the benzene molecule. He knew that it was composed of six hydrogen atoms and six carbon atoms. He also knew that each hydrogen atom bonded with one other atom and that each carbon atom required three partners. He was initially stumped, but then a daydream about a snake eating its own tail delivered to him benzene's circular arrangement of atoms. Machines may crunch through all the possible locations of hydrogen and carbon atoms to determine the structure of benzene, but they won't have daydreams.

A sense of our own specialness leads us to believe that we will retain this advantage into the Digital Age. It's a bit like the confidence of pre-Copernican astronomers that whatever discovery about creation we made, Earth would remain at its center. I argue that this bias in favor of humans is as unsustainable as pre-Copernican geocentrism. It won't benefit the human workers of the Digital Age to mumble mysteriously about improving productivity by implementing the Corbomite Maneuver.

The expected advances that are topics of this book put pressure on our belief in human exceptionalism. Some of the achievements of our species of which we are proudest involve the discovery of patterns in complex phenomena. Albert Einstein detected patterns in the universe invisible to his contemporaries. Some believers in human exceptionalism will airily assert that no machine could ever have come up with the general theory of relativity. But pattern-detection is a *forte* of machine learners. They are built to find patterns in hugely complex sets of data. We may refuse to acclaim them as geniuses when they find an especially obscure pattern in a vast set of data about the genetics and lifestyles of people who acquire autoimmune diseases. But this will not prevent them from finding the patterns. Perhaps judges of the future will accept that genius is as genius does.

Today we look to highly trained humans to advance our understanding of disease. But there's no rule of the universe that requires that treatments for humanity's most feared diseases must be within humanity's powers of inference. Moving into an age in which much of the mind-work about cancer is done by machines could be very good news for our treatment of the disease. We may get therapies beyond the imaginative and logical powers of any human intellect. But it is bad news for our view of human thought and imagination as central to the treatment of our diseases. We will accept the conclusions of the machine learner tasked with treating disease much in the way that the faithful are supposed to accept divine commands. In both cases, ours is not to reason why. We must reflexively make the ceremonial offerings or swallow the pills.

Forward to a Social-Digital Future

My goal in this book is to describe what must be done to preserve human agency in the Digital Age. How are we to avoid writing ourselves out of our own story? I do not claim to possess a crystal ball that permits me to predict every detail of the Digital Age. But we can predict that the Digital Revolution will radically remake work and redirect human agency. That much we cannot change. We can nevertheless influence how it will remake work, how it will redirect human agency. This book presents a vision of future societies whose human citizens have rejected the path of cosmic irrelevance. The preservation of the human contribution will not require a rejection of the technological wonders brought by the Digital Revolution. Rather it will require careful consideration about the domains of human activity that we surrender to the machines.

Societies emerging from the Digital Revolution should be organized around what I call *social-digital economies*. These economies feature two different streams of economic activity, centered on two very different kinds of value. The principal value of the digital economy is efficiency. It focuses on outcomes and cares about means only insofar as they are reflected in outcomes. One process might be more efficient than another because it produces more of a valued product, or produces them more cheaply, or requires fewer raw materials. The principal value of the social economy is humanness. It is founded on a preference for beings with minds like ours, a preference to interact with beings with feelings like ours. We enjoy the company of other members of "the mind club," a phrase I take from psychologists Daniel Wegner and Kurt Gray. Wegner and Gray define the mind club as "that special collection of entities who can think and feel."[11] When we hear that an octopus is conscious we take a special interest in it. We can wonder what it might be like to think octopus thoughts and to feel octopus feelings. Tablet computers are fascinating in many ways, but they lack this kind of appeal for us. We take an extra-special interest in members of the human chapter of the mind club, beings with minds like ours.

This preference for members of our chapter of the mind club operates in the personal domain—it guides our selection of lovers and friends. It also operates in the domain of work. If we think about it, we want our baristas and nurses to have minds like ours too. We will rightly reject the inefficiencies of humans when they stray into parts of the economy that emphasize the skills of the computer. But we should have the courage to reject digital technologies when they trespass on distinctively human activities. We should question the suggestion that advances in artificial intelligence will soon fill our societies with machines that have human feelings and thoughts. Our contract with the machines should be one in which we do the jobs for which feelings matter and they take on many data-intensive tasks for which feelings are irrelevant.

The social-digital economy is a view of humanity's future informed by aspects of our pre-civilized past. It aims to reinstate some of the social aspects of the foraging lifestyle. Efficient digital technologies will shunt humans out of jobs that don't require direct contact with other humans. We will be free to take up jobs in a radically expanded social economy.

This expanded social economy could respond to one of the defining ills of our time—social isolation. Many of the riches of our modern age have come from the denial of our social natures. John Cacioppo, a University of

Chicago psychologist, and William Patrick call humans "obligatorily gregarious."[12] They explain that a zookeeper asked to create an enclosure for *Homo sapiens* would "not house a member of the human family in isolation, any more than you house a member of *Aptenodytes forsteri* (Emperor penguins) in hot desert sand."[13] Cacioppo and Patrick say "As an obligatorily gregarious species, we humans have a need not just to belong in an abstract sense but to actually get together." This obligatory gregariousness is a consequence of evolution. Before the Neolithic Revolution foraging was a human universal. Foraging is an intensely social existence. Foragers live very much in each other's lives. Their shelters tend to be temporary, lacking the permanent walls that separate the nuclear family from outsiders. They share food. Isolation is one of the worst things that can happen—an ostracized forager is almost certainly a dead forager.

We carry the emotional and psychological vestiges of this forager gregariousness into our high-tech times. But many of the resulting needs are unmet. Cacioppo and Patrick say "Western societies have demoted human gregariousness from a necessity to an incidental."[14] They suggest that we see the effects of this demotion in statistics on mental health. Today, isolation causes misery and shortened lifespans. Feelings of social exclusion can manifest as anger or violence.

Technological progress has abetted the demotion of human gregariousness. The less sophisticated technologies of foragers mean that their survival depends on how they get on with other human beings. In his influential book on Americans' increasing retreat from political and social engagement, *Bowling Alone*, the sociologist Robert Putnam observes "In round numbers the evidence suggests that *each additional ten minutes of commuting time cuts involvement in community affairs by 10 percent*—fewer public meetings attended, fewer committees chaired, fewer petitions signed, fewer church services attended, less volunteering, and so on."[15] The commute is a consequence of the car, one of the central wonders of the Second Industrial Revolution. Workers with cars no longer had to live in cramped physical proximity to their places of work. They could spread out into the suburbs and commute to work. At the time of the suburb's invention, it wasn't obvious how much time people would soon be spending sitting alone, wedged into slow-moving rush-hour traffic.

The names we've given to some Digital Revolution technologies suggest that it might restore some of this gregariousness. Social networking technologies are called "social" for a reason. Facebook is all about connecting and sharing. Enhanced connectedness is one of the primary rationales for the Internet. But the varieties of connectedness offered by the Digital Revolution do not give us what foragers get out of their face-to-face, in-your-face, social interactions. Technological mediation makes those connections less direct. It purges many of the foragers' trappings of sociality. A smiley face emoji is not the same as a grin indicating assent to a proposal. There is no opportunity to place a hand on the shoulder of someone whose facial expressions indicate doubt and concern. The accumulation of Facebook friends and Twitter followers brings few of the benefits offered by forager bandmates. When foragers want something difficult done, they must make direct face-to-face connections with bandmates. As they present a plan they interpret facial expressions to determine how likely verbal assurances are to be followed by actual help. Enlisting additional help involves much more than writing an email and wondering "Should I Cc Malika in?" Foragers would not tolerate the anonymized bullying and stalking behaviors that proliferate on the Internet.

We seem to be playing the roles of zoo animals depressed by their unstimulating environments. Among the most important evolved behaviors of animals are those whose purpose is finding things to eat while avoiding things that eat them. Zoos give them ample calories and the security of a cage that isolates them from any natural predator. There are zoo breeding programs for animals that keepers deem especially worthy. The bored pacing of tigers and collapsed dorsal fins of adult male orcas send a clear message of psychological malaise despite these pluses. The technological mediation of our relationships and digital substitutions for human interlocutors seem to short-change us in a similar way.

We shouldn't overstate the implications of this "forward to the past" vision of the human future. There was much that was bad about the lives of pre-Neolithic foragers. They were typically a few failed hunts and a few unproductive days of gathering away from starvation. It would be absurd to say that we should dump our smartphones, cars, and dishwashers and attempt to reinsert ourselves into the ecological niche once occupied by some tens of thousands of pre-Neolithic foragers. For a start, the mathematics don't work. There are, as of early 2018, 7.6 billion humans. Who among us would get to live to enjoy the splendors—and the horrors—of pre-Neolithic foraging? We should be exceedingly grateful for the gifts of technological progress that separate our lives from theirs. But this gratitude should not prevent us from acknowledging a valuable feature of our former foraging existences that we could seek to reinstate. The Digital Revolution offers an unprecedented opportunity to do this.

We won't be making do with the forager's spear and temporary shelter. The digital part of the economy places a premium on the efficiencies brought by powerful digital technologies. We should expect a progressive displacement of human workers from this side of our social-digital economy. We cannot hope to match the efficiencies of the machines in these domains. Human pilots will be unable to compete with the automated flight systems of the future. Machines will perform our keyhole surgeries to a higher standard than any human doctor. We will be free to take up roles in an expanding social economy that replicates some of the social aspects of ancestral human foraging communities. Fabulously efficient digital technologies will be humming away in the background of this recovered gregariousness.

As human workers are displaced from lines of work centered on efficiency, we should be free to take up new varieties of work that meet human social needs. In today's technologically advanced societies, "social worker" is the name of a job that addresses the most extreme harms caused by social isolation and indifference. In a Digital Age centered on a social-digital economy there should be a great diversity of social work. Human social needs are varied and complex. It is most unfortunate that jobs that place humans into direct contact with other humans are foremost among those that our current emphasis on efficiency is causing us to seek to do without. Automated checkouts and customer service AIs are taking the places of workers who deal directly with other human beings. My purpose in this book is to show that machines will always be poor substitutes for humans in roles that involve direct contact with other humans. Here we value connections, however fleeting, between human minds. We care about what's going on in the minds of those who provide services in the social economy. Efficiency is a factor in such interactions, but it is not the only consideration. You want the person with whom you lodged an order for a café latte not to forget to make it. But you value the human interaction that occurs as it is handed over. When a drive for increased efficiency causes us to do without human workers, we leave ourselves ill-prepared for the kinds of digital future that we should be seeking. If we want a truly social Digital Age, then these are the roles that we should be preserving and promoting. We should aim for a future in which machines do much of the heavy lifting and hard calculating but humans find work meeting the many social needs of other humans.

Must this socializing be work? Some argue that we should respond to digital advances by offering humans a universal basic income. Martin Ford argues that a principal role for the humans of the Digital Age will be to go shopping.[16] We will use our shares of the profits generated by the robots to maintain demand for the stuff they make. I am skeptical of this proposal. It overlooks one of the great benefits of the work norm, the idea that our children should grow up expecting to join the work force. Humans may be obligatorily gregarious, but when left to our own devices that gregariousness tends to be parochial. We seek out people we know or people who resemble us in ways that we care about. We fear strangers. Work requires us to get along with strangers. We must cooperate with them to achieve shared goals. Work is part of the success story of the diverse, multi-ethnic societies of the early twenty-first century. Without the social glue of work, some other way will need to be found to prevent the fracturing of our societies into sub-communities defined by ethnicity, religion, and other socially salient traits.

The idea of a social economy brimming with new jobs may seem fanciful. It doesn't seem realistic to suppose that the workers of the near future will be told "The bad news is that we're going to let you go from your job at the supermarket checkout. The good news is that we're creating much more rewarding job for you in the emerging social economy." The social-digital economy is not a prediction. Rather, it's an ideal about how human societies could be in the Digital Age. When Martin Luther King Jr. intoned "I have a dream" it was not appropriate to respond "Yeah right, dream on." We should reject overconfident forecasts of the societies of the Digital Age. We should nevertheless approach the Digital Age with an attitude of empowered uncertainty. The social-digital economy is an ideal sufficiently attractive to be worth fighting for. There are many ways in which we could fail to realize the ideal of the social-digital economy. We could resign ourselves to one of the many dystopian visions in which most of the humans of the Digital Age have existences that are both pointless and impoverished. In some of these visions, all the wealth generated by digital machines goes to the few who own them. Reflection on the increasing inequality of our age does present this possible future as the path of least resistance. Alternatively, we could seek to create a Digital Age in which we are surrounded by fabulous digital technologies but still manage to enjoy intensely social existences. The route toward the social-digital economy will not be easy. It will require tough choices. We must muster the collective will to reject some of the superficially appealing offers that digital technologies present to us. Were I to place a wager, I would bet against Digital Age societies centered on social-digital economies. I would also bet against our collectively finding an adequate response to human-caused climate change. I find it hard to see in the responses that we have collectively managed thus far, anything indicative of a response sufficient to prevent an ecological catastrophe. But when it comes to the future of the human species I'm not a betting man. In both the case of climate change and the threat from the Digital Revolution to human agency the rewards of success and penalties for failure are so high that they call for our greatest efforts. We must do all we can to create societies centered on social-digital economies.

A Note on Philosophical Method

I am a philosopher and I treat questions about human agency in the Digital Age as philosophical problems. However, my attitude toward philosophical problems is somewhat different from those taken in other philosophy books.

I view philosophy as at its best when it plays an integrative role. An understanding of the threat to human agency posed by the Digital Revolution should draw on many different sources of information. This book makes use of insights from experts on digital technologies and large-scale technological change, social psychologists, economists, evolutionary biologists, and philosophers of mind. Philosophers have the intellectual skills to integrate all these different kinds of information into a coherent approach to the Digital Revolution's social transformations. The questions addressed by philosophers are noteworthy for their variety. Philosophers are archetypal academic generalists. We generate no data. When we address questions including the nature of art, the existence or nonexistence of subatomic particles, and the possibility of a just state, we are frequently challenged to find proper assessments of the significance of ideas from outside of philosophy and to integrate them with other ideas with different provenances. Philosophers are, in effect, informational and theoretical brokers facilitating exchanges in ideas between different academic disciplines that don't normally find themselves in contact.

One way to go wrong in dealing with problems like the human consequences of the Digital Revolution is to fail to properly acknowledge an important source of information or to overstate the significance of a favored type of information. In chapter 1, I criticize the economist Robert Gordon because his forecasts rely too heavily on historical economic data and pay insufficient attention to the kinds of trends affecting digital technologies. Gordon displays excellent understanding of the economic data he has gathered but he offers unreliable advice about the future because he does not understand the significance of expected developments in artificial intelligence.

There is another important sense in which this is a philosophy book. My proposal of a social-digital economy makes essential use of the insights of philosophers of mind on the nature of phenomenal consciousness—the "what it's like" aspect of human thought. I argue that we should not expect machines to meet our need to interact with others who have feelings like ours. My integrative approach leads me to treat philosophical expert evidence about phenomenal consciousness in the same kind of way that I treat the expert evidence of economists, evolutionary biologists, technologists, and social psychologists. We seek an appreciation of the relevance and relationships between of all these different sources of information if we are to achieve an understanding of the human consequences of the Digital Revolution. When I dispute the claims of economists about the future of work I do not pretend to advance our understanding of economic theory. This doesn't prevent me from disputing the detail of their claims about work in the Digital Age. Nor does my use of ideas from the philosophy of mind aspire to solve deep philosophical problems about the nature of phenomenal consciousness. Rather, I make claims about rational ways for us, as humans, to respond to philosophical doubts about whether machines can have feelings like ours. Instead of solving deep problems in the philosophy of mind, I show how philosophical ideas should influence our approach to the Digital Revolution.

An Outline of the Book

Chapter 1 introduces the long view of the Digital Revolution. I propose to locate it alongside other technological revolutions rightly counted as turning points in human history—the Neolithic and Industrial Revolutions.

The second decade of the twenty-first century is not the ideal time to be making assessments of overall historical significance. We are at the height of excitement about all things digital and subject to a tendency to overstate the computer's significance. The economist Robert Gordon argues that the Digital Revolution will not live up to the expectations of its enthusiasts. He compares the Digital Revolution with the Second Industrial Revolution, which was centered on electricity and the internal combustion engine. Gordon says that advances derived from the Second Industrial Revolution "covered virtually the entire span of human wants and needs, including food, clothing, housing, transportation, entertainment, communication, information, health, medicine, and working conditions."[17] The effects of the Digital Revolution are restricted chiefly to entertainment and information and communication technology. They lack the broad human and economic significance of the earlier technological revolution. I argue that Gordon sells the Digital Revolution short. The expected application of artificial intelligence to vast quantities of data points to important impacts beyond entertainment and information and communication technology.

Artificial intelligence is the focus of chapter 2. The goal of work in artificial intelligence seems easy to state—it involves the attempt to build a machine with a mind. I suggest that work in AI has developed a split personality. We can distinguish a philosophical motivation directed at creating machines with minds from a pragmatic motivation that aims to build machines that do mind work—the things that humans use their minds to do. The philosophical motivation was initially described by Alan Turing, AI's founding genius. It is an excellent premise for a movie. But the principal focus of work on AI in the early decades of the twenty-first century is pragmatic. Pragmatists are now making machines that do mind work better than humans. Turing's desire to create a machine capable of authentic thoughts becomes a hobby project when placed alongside the pragmatic goal of building machines capable of exploiting the wealth inherent in data and solving our most challenging problems.

Chapter 3 switches focus from artificial intelligences to the data that is the focus of their mind work. It defends the wisdom in the popular Internet saying "Data is the new oil."[18] Data is the Digital Revolution's defining form of wealth. Corporations' holdings of data are coming to matter more in assessments of their market value than do their holdings in varieties of wealth specific to earlier technological revolutions, such as land or oil.

I argue that some of us are slower on the uptake in our understanding of this new form of wealth and this leaves us at a disadvantage when dealing with those who have better understanding of data's value. We renounce control over our data to Google, Facebook, and 23andMe, much in the way early twentieth-century Texas farmers happily accepted paltry sums for the right to prospect for oil on land that wasn't of much use for ranching. I consider the relevance in the long view of the aphorism "information wants to be free." Stewart Brand, Kevin Kelly, and Jeremy Rifkin expect a future in which data resists attempts to control and limit access to it. I suggest that we assess such claims in the context of a broader political and economic context in which some people do very well out of asserting exclusive control over data. I consider Jaron Lanier's suggestion that we should respond by charging a fee—a micropayment—for the privilege of using our data. Streams of micropayments would flow back to the originators of Internet content. I challenge the practicality of this idea.

Chapter 4 presents the Digital Revolution's threat to human agency. Considered at its most prosaic, the threat to human agency is a threat to our jobs. The superlative mind workers of the Digital Revolution reduce the economic value of human agency. Why pay a human to do a job that can be done better and more cheaply by a machine? We have many precedents for the devaluation of human agency by technological progress. I consider the inductive case for optimism about the Digital Revolution advanced by many economists and commentators on technology. It's difficult to imagine the jobs that the Digital Revolution will create. But, if the past is a guide, we can assume that new jobs will be created. The grandchildren of accountants and waiters will be relieved that they don't have to make their livings inserting numbers into spreadsheets or waiting on tables. I reject this optimism. The Digital Revolution brings new economic roles that could be filled by humans. But the protean powers of digital technologies should promptly eliminate any new jobs. The money paid to human workers creates a powerful motivation to build cheaper and more efficient digital substitutes. The higher the economic value of a new role, the stronger the incentive to automate it.

We have a debate between optimists and pessimists about the Digital Age. I argue that we should approach the challenge of automation to work as pessimists. Optimism is a therapeutic way for us as individuals to confront life's challenges, but it is a bad way for us to collectively confront the challenges of the Digital Revolution. Paying more attention to the forecasts of the pessimists offers better insurance against an uncertain future than does the feel-good message of the economists' inductive optimism.

Chapters 5 and 6 explore a possible haven for the human workers of the Digital Age. Thus far, we have compared humans and machines in terms of efficiency. An interest in efficiency focuses on outcomes. We focus on means only insofar as they are reflected in outcomes. An interest in increasing efficiency should lead to the progressive elimination of humans from the economy. An alternative value is humanness. The value of humanness directs us to prefer human means. Humanness matters a great deal in our most important relationships. There are many science fiction explorations of scenarios in which romantic partners are replaced by machines that efficiently perform all the actions related to romance but whose mental lives we doubt. Even if not fully acknowledged, we carry this interest in human experiences into the domain of work. We assume that the doctors who tend to our wounds, baristas who make our espressos, and politicians who make decisions about our societies' minimum pay rates have mental lives like our own. They have feelings like our own. They are members of the human chapter of the mind club. We care about efficiency in these domains too. It's bad when your barista forgets your order or when your nurse injects the wrong medication into your arm. But we tend to respond to inefficiencies in ways that preserve human contributions. When we learn that human nurses sometimes give patients the wrong medication we don't seek to replace them with machines. Rather we supplement them with machines that correct these errors. We seek to eliminate the error—thereby improving efficiency—while preserving the distinctively human contribution.

We should be heading toward a social-digital economy. This bifurcated economy should see the progressive replacement of human workers in nonsocial domains. Machines will fly our airplanes and perform our keyhole surgeries. Humans will retain an ongoing superiority in roles for which social contact with other human beings is central. In many cases, they will be assisted by powerful digital technologies. But we will justifiably view the contributions of these digital technologies as having a significance secondary to human contributions. Our gratitude for a service we have received will go to the human part of the team that performed it rather than to the machine.

It's one thing to assert that the social side of the social-digital economy contains jobs that are best performed by humans. Can we be confident that there will be enough such jobs to absorb the many workers excluded from jobs that do not exploit our social abilities? I suggest that there *should* be. I don't predict that there *will* be. If we don't create these roles they won't exist. One of the great problems for the technologically advanced societies of the early twenty-first century is social isolation. Humans are intensely social creatures. The conditions of our evolution involved constant contact with other humans. But the environments we have recently created tend to isolate us from each other. Lonely humans are more miserable and die younger. A social-digital economy would create new jobs defined around our social natures. Humans will find employment in roles designed to meet the many social needs of other humans.

Is a social-digital economy more than just wishful thinking? Chapter 7 offers advice about how to understand this call for the creation of a social-digital economy fit to carry humans into the Digital Age. I do not offer the social-digital economy as a prediction. The path of least resistance directs our species toward a dystopia in which the members of a small elite own all the machines and hence almost all the wealth. The rest of us are subject to a poverty void of meaning. I offer the social-digital economy as an ideal that we can strive to realize even as we recognize that the odds are stacked against it.

One competing ideal comes in the form of a universal basic income (UBI). Perhaps the ideal of an intensely social Digital Age economy is appealing. But must that socializing be work? Some commentators look forward to a future without work. They call for a universal basic income that would redistribute to the workless some of the wealth generated by increasingly efficient machines. This book embraces a future *with* work. Good work provides social benefits and is therapeutic. We should complain about work that is dirty, degrading, and dull but not about work itself. The new jobs on the social side of the social-digital economy aren't required to have the unpleasantness of many of today's jobs. They should engage our social natures and they should also lack the objectionable features of many of the jobs most imminently threatened by increasingly efficient machines.

Chapter 8 offers some practical pointers about how to make the Digital Age more human. What can we do now to stick up for our humanity in an age of superlative digital technologies?

Chapter 9 brings together some intellectual threads. I express a hope that humanity's next age will be named not for its dominant technological package but instead to affirm the social nature of our collective existences. Humanity could exit the late Industrial Age, pass through the Digital Age, and enter a Social Age.

1 Is the Digital Revolution the Next Big Thing?

When we take the long view of the digital computer, we ask what might make it a turning point in human history. We look beyond the buzz that accompanies the latest iPhone or the release of the Oculus Rift virtual reality headset. We know that these feelings of excitement tend to pass. We understand that our spectacular digital devices will seem to future people as Robert Stephenson's 1829 "Rocket" train traveling at 40 kilometers per hour seems to passengers on today's Shinkansens and Trains à Grande Vitesse regularly exceeding 300 kilometers per hour. The adjectives that we excitedly apply to our latest digital gadgets are unlikely to capture aspects of them that change the course of human history. What might make the Digital Revolution similar to other acknowledged turning points in history?

This book presents the introduction and spread of the digital computer and allied technologies as a technological revolution—the Digital Revolution. I take the idea of technological revolutions as drivers of history from the work of V. Gordon Childe, an iconoclastic early twentieth-century Australian anthropologist.[1] Childe introduced the concept of the Neolithic Revolution to describe the diffuse and stop-start yet inexorable move from foraging to farming. It signaled the end of the Mesolithic Age—or middle stone age—a time when the foraging way of life was a human universal. The Neolithic Revolution's earliest verified location was the Fertile Crescent—a region running from the areas around the Nile River to modern-day Iran and Turkey—of 12,000 years ago. But the transition from foraging to farming occurred independently in a variety of other locations. The Neolithic Revolution brought ploughs and other technologies that permitted farmers to make deliberate productive use of the fertility inherent in the ground under their feet. Crops were planted. Animals were husbanded.

The Neolithic Revolution radically transformed both individual human lives and human societies. Land that once supported only a small number of foragers could support many more farmers. The new sedentary lifestyle prompted radical changes in the way humans organized themselves into societies.

Central to Childe's discussion of technological revolutions was the idea of a technological package. The Neolithic package contained technologies related in some significant way to agriculture. The title of his 1936 book *Man Makes Himself* sums up his view of the human significance of the introduction of this technological package. The Neolithic package includes crops, domesticated animals, stone agricultural tools, pottery, and the permanent dwellings required by sedentary peoples. Fast-forwarding several millennia, we find the distinctive industrial package that brought the Industrial Revolution. The steam engine, the factory, and new production methods were elements of this package.

The notion of a technological package suggests a special relationship between constituent technologies, a *functional interdependence*. The technologies that comprise a package relate to each other in significant ways.[2] Improvements in one technology belonging to a technological package suggest or motivate improvement of other components of the package. The cultivation of crops brought a more settled existence. This motivated the development of more substantial and permanent dwellings. A society that plants crops tends to benefit greatly from the invention of pottery containers. The technologies that comprise a technological package hang together as a coherent mutually supporting and interdependent whole.

Childe was guilty of oversimplification. As the archeologist Steve Mithen explains, Childe supposed that the Neolithic package was "always acquired as a single, indivisible whole."[3] This seems not to have been the case. People in different places got along for some time with incomplete Neolithic packages. There was no *2001: A Space Odyssey* event in which a black box descended on a band of Mesolithic foragers and imparted to them the five or so basic technologies Childe presented as comprising the Neolithic package. Nor did any Moses-like Big Man of a forager band ever go up a mountain and return with knowledge of crops, animal husbandry, pottery, and the principles of stone masonry.

Some foragers acquired pottery and showed little disposition to acquire other elements of the Neolithic package. Some societies made significant headway into agriculture without the multiple benefits of pottery.[4] But these alternative trajectories of development do not prevent pottery from joining other elements of the Neolithic package as part of a functionally coherent whole. Pottery-making techniques enable the creation of items with great spiritual significance for foragers. But their impact on individuals and societies is so much greater when combined with crops. There is, to use a word whose impact has been blunted by overuse, a synergy between pottery-making and the cultivation of cereals. A pottery container filled with grain is something significantly more than just a piece of pottery. Combining pottery with crops significantly boosts the benefits conferred by each.

The networked computer is the Digital Revolution's technological package. As we saw with pottery and farming, the technologies that comprise this package are separable. Telephones were networked before there were computers. The invention of the digital computer that inaugurated the Digital Revolution preceded the insight that it was possible to network them. Computers deliver great benefits to a society without having to be networked. Some societies today would like to bestow powerful computational abilities on workers in economically or politically significant industries while limiting the social disruption that they view as emerging from the unrestricted networking of those computing devices. The rulers of North Korea want selected workers to be able to calculate the yields of nuclear bombs, and to use their Kwangmyong "walled garden" national intranet, but they don't want them to go on Facebook. The fact that networking and computers are part of a mutually interdependent package of technologies suggests that unusual conditions are likely to be required to prevent the one from bringing the other. Computers tend to bring networking with them. When combined, the technologies bring greater benefits than the sum of the benefits each produces independently. Unusual conditions, such as those containing vigilant and repressive police forces, may prevent the digital computer from bringing the benefits of networking. The synergistic relations between the technologies that comprise a package do not prevent variations in local conditions from causing some elements of the package to be present in locations where others are absent. But this should not prevent us from acknowledging elements of a technological package as mutually interdependent. They tend to go together. When found with other members of a package, technologies amplify the impacts of other elements.

Childe's notion of a technological revolution suggests the idea of a technological age. The Neolithic Revolution inaugurated the Neolithic Age. The Industrial Revolution inaugurated an Industrial Age. There is an important sense in which the package of technologies introduced by a technological revolution sets some of the ground rules for the age it initiates. Advances during a technological age tend not to have the rule-making significance of the introduction of a new technological package.

Now is not the ideal time to make assessments of comparative historical significance. We currently find ourselves at a stage of peak excitement about the life-altering power of the networked computer. When the same people who brought us the iPhone and the Google search engine talk about driverless cars and using machine learning to cure cancer we seem licensed to treat these possibilities with a credulity that didn't seem appropriate for talk by 1960s futurists about colonies on the planet Mars. We know that they can do this kind of stuff. The expression "and pigs will fly" no longer seems to describe a complete absurdity so long as we append the word "digitally." This excitement has a distorting effect on our assessments of the historical significance of the Digital Revolution. We should ask ourselves whether today's assessments of the earth-shaking impact of the latest digital technologies are akin to a teenager's pronouncement of the band that currently captures her interest as "the best band ever!" a label promptly appropriated by next week's musical obsession. Our current historical location at the epicenter of excitement about everything digital is far from the ideal standpoint for sober comparative judgments about the historical significance of the Digital Revolution. But reasoned assessments of comparative historical significance are possible. We can speculate about the considered judgments of posterity. How will people who view Apple's product launches as historical marketing curiosities assess the significance of the Digital Revolution?

Will the Digital Revolution Fizzle?

Suppose that we accept Childe's general framework. Technological revolutions centered on technological packages set the ground rules for societies of the subsequent technological age. The Neolithic and Industrial Revolutions were such events. Does the Digital Revolution belong with them? It's useful to place current enthusiasm about the Digital Revolution in the context of a recent skeptical assessment of it advanced by Northwestern

University economist Robert Gordon. Gordon argues that the Digital Revolution is no turning point in human history and he seems to have the numbers to back him up.

Gordon argues that the Digital Revolution's boosters overstate the human and economic significance of the networked digital computer.[5] He offers this assessment in the context of an impressive history of the economic consequences of technological innovation. Gordon's 2016 book, *The Rise and Fall of American Growth,* makes the case that the century from 1870 to 1970 was unique in human history for its rapid improvement of human standards of living and rates of economic growth. According to Gordon, this growth occurred as the result of a range of unrepeatable technological innovations. These innovations radically changed life at home and at work. Gordon says "The economic revolution of 1870 to 1970 was unique in human history, unrepeatable because so many of its achievements could happen only once."[6]

Gordon's language indicates his modest opinion of the networked computer. In Gordon's terminology, the Digital Revolution becomes the Third Industrial Revolution—or, more belittlingly, IR #3. In Gordon's long view, there was the First Industrial Revolution (IR #1), "based on the steam engine and its offshoots particularly the railroad, steamships, and the shift from wood to iron and steel," and the Second Industrial Revolution (IR #2), which "reflected the effects of inventions of the late nineteenth century—particularly electricity and the internal combustion engine."[7] For Gordon, IR #3 is the third of the industrial revolutions not only in temporal order, but also in overall significance. IR #3 is a disappointment after IR #2's epic effects on human well-being and economic growth.

Gordon summarizes the human significance of the one-off improvements of IR #2 at home under the heading of networking. By "networking" Gordon means something whose human significance was much more expansive than connecting computers to each other. It describes a radical transformation of home life in the hundred years beginning at 1870. Gordon says "the 1870 house was isolated from the rest of the world, but 1940 houses were 'networked,' most having the five connections of electricity, gas, telephone, water, and sewer."[8] There have been many improvements to the manner of networking but none of these match the significance of the initial connection to electricity, gas, telephone, water, and sewer. The hundred-year period that is Gordon's principal focus took many people from nothing to something. The years after 1970 have, in general, taken us from something to something a bit better. The human and economic magnitude of the leap from houses lit by oil lamps to houses lit by electricity is greater than the transition from houses lit by electricity to houses with safer and more reliable supplies of electricity. None of the many improvements to the electrical grid seem to match the significance to householders of their homes' initial connection.

There were unrepeatable innovations outside of the home too. Cars replaced horses bringing suburbs into existence. Passenger jets promptly and cheaply transported people over distances that could formerly be traversed only with great effort and at great cost. The cars and jets of 2018 are superior to those of 1970 but the human and economic significance of these 48 years of improvement cannot match the significance of the original introduction of the car and the passenger jet. This pattern of technological transformation is evident in the workplace too. There was a general shift form hard outdoor labor to work in air-conditioned offices. The experiences of workers have changed since they relocated to a temperature-controlled indoors. But none of the changes since 1970 match the significance of the initial move inside.

Gordon proposes that the one-off nature of these advances explains a plateauing of progress since 1970. He maintains that digital technologies have failed to correct a deceleration of economic growth noticeable since 1970. He allows that there was a ten-year period of rapid progress in productivity associated with IR #3 from 1994 to 2004 "when the invention of the Internet, web browsing, search engines, and e-commerce produced a pervasive change in every aspect of business practice."[9] But after this brief surge, growth reverted to a more pedestrian pace, far from that characteristic of growth unleashed by IR #2. Says Gordon: "Though there has been continuous innovation since 1970, it has been less broad in its scope than before, focused on entertainment and information and communication technology (ICT)...." He continues "Like IR #2, it achieved revolutionary change but in a relatively narrower sphere of human activity." While IR #2 "covered virtually the entire span of human wants and needs, including food, clothing, housing, transportation, entertainment, communication, information, health, medicine, and working conditions," the effects of IR #3 are restricted to "only a few of these dimensions, in particular entertainment, communications, and information."[10] In effect, when we compare IR #1 and

IR #3, we are comparing railroads and internal combustion engines with posting status updates to your Facebook friends, targeting advertisements for holidays in Fiji to those who Google search "three-star hotel on Denarau Island," and streaming high-definition *The Walking Dead* episodes.

The problem for IR #3 lies not so much in any limitations inherent in digital technologies themselves. Gordon acknowledges that some of these are becoming more powerful at an exponential pace. The deceleration in the economic effects of technological progress seems to say something about us. The technologies that grew out of IR #2 satisfied many longstanding human needs and wants, leaving comparatively little scope for the shiny tech advances of IR #3 to measurably improve the human condition. Someone today can load apps to their smartphone that warn them when their fridge's supply of milk is getting low, but this piece of technological wizardry cannot match the significance of the original introduction of the domestic electric refrigerator. When data flows down fiber-optic cables it brings especially crisp images to televisions. These images are appreciably better than those they replaced. Gordon allows that "People could now watch a station devoted entirely to music videos or, if they were willing to pay extra for premium cable, could see movies on HBO, free from advertising interruption, twenty-four hours a day."[11] Home Box Office did well off its "It's Not TV, It's HBO" slogan. Really, however, it is TV, just a bit better. A 2016 episode of HBO's fantasy epic *Game of Thrones* with its CGI dragons, sprawling battle scenes, and occasional nudity seems to differ greatly from a fully-clothed, black-and-white 1955 episode of CBS's western *Gunsmoke*, but they satisfy essentially similar viewer needs. There's certainly a big difference between *Gunsmoke* and *Game of Thrones* and anything happening in the living rooms of 1894. The years 1870 to 1970 took us from "household drudgery, darkness, isolation, and early death" to a time when human fundamentals differed little from today.[12]

This is not to say that once we got the refrigerator, the color television, and the passenger jet that we pronounced ourselves fully satisfied. But perhaps it is to say that that portion of human needs that are readily accessible by technological progress has largely been sated. We must look beyond the domain of technological innovation to satisfy extant human needs. Gordon identifies inequality as a "headwind" to economic growth.[13] Inequality is partly a matter of how access to technologies is shared out rather than facts about the technologies themselves. Technological advances often lead to more unequal outcomes. A society in which everyone travels by horse-drawn wagon is more equal, in respect of transport, than one in which half of its members get to travel in automobiles.

One of the impressive things about Gordon's book is the wealth of data on which he founds his claims about the impact of technological change on economic growth. Philosophers are often reduced to speculation to support their suggestions about trends and future directions. Gordon is the beneficiary of a large amount of data about economic growth over the miraculous century 1870 to 1970 and thereafter. Gordon's measure of the economic impact of technological progress is Total Factor Productivity (TFP). Gordon defines TFP as "output divided by a weighted average of labor and capital input."[14] Holding fixed the contributions of labor and capital permits us to isolate other influences on the growth of an economy's output. Technological improvements are high on the list of reasons why a later time's output is higher than an earlier time's. When Gordon complains about a slackening of growth since 2004, he is doing more than reporting on a general impression that things have gotten better in the years since 2004 but at a considerably slower pace than the century from 1870 to 1970. He has the data. But an overreliance on data can mislead about the future if there is good reason to think that data up to the present day fail to adequately describe some impending change.

The Magic Combination of Artificial Intelligence and Data

Why might the recent past of digital innovation fail to predict its future effects on economic growth and human well-being? Some of the most eloquent popularizers of the Digital Revolution point to the exponential pace of improvement of some of its technologies.[15] We observe exponential improvement in changes to integrated circuits, Internet bandwidth, and the storage of data on magnetic hard disks, to name just a few. What separates the past of exponential improvement from its future is that we are about to go through a significant transition. Those who write about exponential technological change are especially impressed by the hockey-stick shape of the graphs that describe it. After a slow and unspectacular beginning, the line of exponential growth goes near vertical. Erik Brynjolfsson and Andrew McAfee call the transition from slow to rapid growth the "inflection point."[16] They join other advocates of exponential improvement including the *New York Times* journalist Thomas Friedman in suggesting that a wide range of digital technologies are entering their inflection points.[17] Friedman's preferred verb for this change is "explode" and his favored adjective "explosive." Gordon's data could be misleading if he is sampling from the unimpressive, slow growth phase of IR #3's curve. A general loss of business confidence in the years after the bursting of the Dot-com bubble in the early 2000s might explain the dip in economic growth reported by Gordon. But if digital technologies are on the point of entering their inflection points, then their impacts on the economy could become increasingly emphatic.

In what follows, I prefer an explanation that points not to accelerating changes to the power of digital hardware but rather to what we are increasingly doing with this hardware. The difference between the IR #3 up until now and the results it will soon produce lies in the data that digital machines are gathering and processing. The magic ingredient that we are increasingly mixing with this data is artificial intelligence. Digital machines are becoming intelligent. This intelligence is permitting them to do mind work—the work that humans traditionally do with our minds.

Remember that Gordon suggests that a key difference between IR #2 and IR #3 lies in the breadth of human wants and needs that they address. He says that IR #2 "covered virtually the entire span of human wants and needs, including food, clothing, housing, transportation, entertainment, communication, information, health, medicine, and working conditions." IR #3 has thus far satisfied our needs and wants in "only a few of these dimensions, in particular entertainment, communications, and information."[18] The miraculous combination of data and artificial intelligence suggests that IR #3 will begin to satisfy needs and wants outside of those dimensions.

Gordon's diagnosis does seem to describe the economic impacts of the networked digital computer up until now. Google's advertising business comprises the bulk of the revenue of its parent company, Alphabet.[19] The idea that it is possible to make money from advertising is not new. People in the advertising industry have long known that messages generate more sales when they are directed at people more likely to buy. Google's AdSense and AdWords permit advertisements to be very precisely targeted. They give advertisers an unprecedented access to the minds of potential buyers. They seem to be a significant step toward what technology writer Tim Wu calls "advertising's holy grail"—"pitches so aptly keyed to one's interest that they would be as welcome as morning sunshine."[20] It's easy to see why Toyota might be prepared to pay quite a lot of money to display its ads on the web browsers of people who type "best compact car" into Google. It's also easy to see how people who type those words might have a different experience of Toyota's advertising than they have of most Internet sales pitches.

Gordon's pessimism about the Digital Revolution would be justified if targeted advertising, email, online news, and streamed movies were basically it.[21] But there is a difference between a description of current uses of data and uses we can confidently predict. The entrepreneur and data scientist Jeff Hammerbacher offers the following complaint about the dominance of advertising: "The best minds of my generation are thinking about how to make people click ads. That sucks."[22] Hammerbacher's complaint is best interpreted as impatience about the delay in getting on to explore the digital package's more ennobling applications. He should draw confidence from the realization that clicking on ads is just the beginning for our exploitation of data.

Yogi Berra warned "It's tough to make predictions, especially about the future." The following discussions about transportation and health involve predictions about the future. They present conjectures that could turn out to be false. They are, nevertheless, reasonable extrapolations of current or expected digital technologies. Artificial intelligence is the Digital Revolution's killer app. There have been some vivid demonstrations of the power of AI. In the 1990s, chess computers went rapidly from being able to beat amateurs while being far inferior to the best human players, to being far superior to them. In 2011, the IBM computer Watson sorted through four terabytes of data, including the entirety of Wikipedia, to beat two human *Jeopardy!* champions. AlphaGo, an application of Google's DeepMind AI, was fed data on 30 million Go board positions from 160,000 real-life games.[23] Go is an exceedingly demanding game once thought to be an insurmountable challenge for computers. In 2016, AlphaGo won a match against Lee Sedol, one of the world's top-ranked human players.

Victories in chess, *Jeopardy!*, and Go suggest a wide range of future achievements for AI. Artificial intelligence conveys the transformative power of data into domains of human experience that benefitted from IR #2, but are thus far relatively untouched by IR #3. Remember Gordon's list of aspects of human life thus far comparatively undisturbed by the networked computer, including food, clothing, housing, transportation, health, medicine, and working conditions. In the following pages, I explore how the combination of data and AI should be expected to transform transportation, health, and medicine. Hammerbacher's complaint about any bias toward ad clicking will no longer apply.

How AI Could Transform Transportation

Globally, cars kill about 1.2 million people every year.[24] Compare this figure with the 28,328 deaths attributed to terrorism in 2015.[25] Like soldiers in their third year of seemingly interminable trench warfare, we have inured ourselves to the mayhem on our roads. We've come to accept these deaths as the price for convenient commutes to work and visits to out-of-town family. Driverless cars promise safety improvements unachievable by advances in driver education programs or by grafting additional safety features onto human-driven cars. Somewhat depressingly, human drivers respond to sophisticated driver assist technologies by permitting their eyes to wander from the road more frequently. This was horridly borne out in a July 2016 fatal crash in which a driver placed his Model S Tesla into auto-pilot mode and took the opportunity to watch a Harry Potter movie.[26] For those seeking to make a serious dent on that 1.2 million figure, the best option is to eliminate the human altogether. We should convert ourselves from dangerously distractible drivers to full-time passengers whose control over our cars is limited to linking our smartphones to their computers and speaking the names of our destinations.

Today's experimental driverless cars combine a variety of sensors. They detect nearby objects with LIDAR (Light Detection And Ranging), a system that bounces a laser at nearby objects and analyses its reflections. A radar system focuses especially on fast moving nearby objects by bouncing radio waves off them. To these are added sonar systems that bounce sounds off nearby objects and detect their echoes, and optical cameras. The cars make longer-range plans with the help of mapping software such as Google Maps. This provides a driverless car with satellite imagery, street maps, and real-time updates on traffic conditions.

These sensors generate a great deal of data, which would be of limited value without the capacity to analyze it. In their book, *Driverless: Intelligent Cars and the Road Ahead,* Hod Lipson and Melba Kurman explain how machine learning allows driverless cars to make significantly better use of data than could a human driver fed edited highlights on a dashboard display. Driverless cars engage in a "deep learning" that is radical departure from the traditional AI strategy of trying to program in rules covering all the situations that a car might encounter. The cars are more trained than programmed. The training of a driverless car begins with laborious Darwinian process in which responses that conform to the assessments of human trainers—for example, noticing and avoiding a pedestrian—are preserved, and responses that fail to conform to them are discontinued. Lipson and Kurman describe the decreasing importance of human trainers:

> At some point, the deep-learning software will reach the point where it can guide a driverless car on its own, enabling the car to drive alone and collect a steady stream of new training data as it goes. The new data will be applied to train the deep-learning software to reach even higher levels of accuracy in recognizing objects, further improving its performance. As cars' guiding software becomes even more capable, more driverless cars can be dispatched to the streets, collecting yet more training data.[27]

The longer a driverless car is on the road, the more data it accumulates and the better it handles the road. "Fleet learning" further expands the gap between human drivers and driverless cars by transmitting the results of a particular car's training to all the cars in a fleet. It's as if an irascible parent could just transfer into the mind of their child learner all their many years of driving wisdom without the need for frequent expostulations of "Did you notice that stop sign?!" There are depressing patterns in the fatal distractions and overconfident passing maneuvers of human drivers. The mistakes of driverless cars tend not to be repeated. It is thought that the Tesla autopilot in the crash mentioned above failed to distinguish the whiteness of an 18-wheel truck and trailer crossing the highway from the bright white sky of a spring day. It's unlikely that any Tesla car will make that specific mistake again. Lipson and Kurman present the benefits of fleet learning in the following way.

> As cars pool their driving "experience" in the form of data, each car will benefit from the combined experiences of all other cars. Within a few years, the operating system that guides a driverless car will accumulate a driving experience equivalent to more than a thousand human lifetimes.[28]

Robert Gordon is skeptical about driverless cars. In his dismissal of projected IR #3 developments he awards last place to driverless cars, on a list that includes medical and pharmaceutical advances, small robots and 3D printing, and Big Data and artificial intelligence. Gordon offers two lines of criticism. First, epoch-making though the driverless car may seem to us, it is a comparatively minor tweaking of the hugely significant IR #2 advance of the car. Gordon says "This category of future progress is demoted to last place because it offers benefits that are minor compared to the invention of the car itself or the improvements in safety that have created a tenfold improvement in fatalities per vehicle mile since 1950." He is also unimpressed by the potential economic impacts of driverless cars; he says "The additions to consumer surplus of being able to commute without driving are relatively minor. Instead of listening to the current panoply of options, including Bluetooth phone calls, radio news, or Internet-provided music, drivers will be able to look at a computer screen or their smartphones, read a book, or keep up with their e-mail."[29]

Gordon is unwarrantedly pessimistic about the economic consequences of the driverless car. Currently a huge percentage of the space in our cities is given over to cars. A KPMG survey revealed that half of today's car owners do not expect to own a car when driverless technology is fully realized.[30] As human drivers are eliminated we should see a steep reduction in the number of cars congesting our cities. Many people will liberate themselves from what former generations viewed as the quintessence of liberation. There will be no more frustrated searches for parking. The car that ferried you to your downtown appointment will simply return to its base. Gordon grants that this could have "positive effects on quality of life if not on productivity growth."[31] But the potential economic benefits that accompany these quality of life effects could be considerable. Driverless cars could revitalize city centers currently scarred by crisscrossing highways that transport workers between their downtown jobs and their suburban homes in stressful rush hour traffic. People could frequent businesses that offer a wider range of services than take out coffee and fast food.[32]

The second theme of Gordon's skepticism about driverless technology concerns whether it can be realized. Gordon says "The enthusiasm of techno-optimists for driverless cars leaves numerous issues unanswered."[33] Current driverless prototypes encounter situations that flummox them. They aren't particularly good at deciding when it's safe to pass on a two-lane road. There are glitches in the software that controls current voice-activated control systems. And so on.

David Autor conforms to the theme of economists underwhelmed by digital innovation. He says of machine learning's apparent capacity to find useful patterns in large amounts of data:

> My general observation is that the tools are inconsistent: uncannily accurate at times; typically only so-so; and occasionally unfathomable.... IBM's Watson computer famously triumphed in the trivia game of Jeopardy against champion human opponents. Yet Watson also produced a spectacularly incorrect answer during its winning match. Under the category of US Cities, the question was, "Its largest airport was named for a World War II hero; its second largest, for a World War II battle." Watson's proposed answer was Toronto, a city in Canada.[34]

Autor's crack about Watson is somewhat mean spirited. Even the most commanding tennis performance by Serena Williams includes some bad shots. We should consider Watson's geographical confusion in a context that includes the human quiz contestant who, when asked "How many kings of England have been called Henry?" ventured "Well, I know Henry VIII. So, um, three?"[35]

Autor acknowledges the debates about the prospects for machine learning. He says "Some researchers expect that as computing power rises and training databases grow, the brute force machine learning approach will approach or exceed human capabilities. Others suspect that machine learning will only ever 'get it right' on average, while missing many of the most important and informative exceptions."[36] But Autor's summation of machine learning favors the pessimists. He notes that many objects are defined by their purposes—a chair is something designed for humans to sit on and that these purposes present "fundamental problems" for machine learners even if they have large stores of data from which to learn. Current machine learners struggle to distinguish chairs from objects with similar dimensions that no human would try to sit on. Autor presents his estimation of the magnitude of the challenge confronting machine learning by citing Carl Sagan's observation: "If you wish to make an apple pie from scratch, you must first invent the universe."[37] Difficult!

It's clear that today's most powerful machine minds can do stupid things and that machine learning faces difficult challenges. The question is how these challenges feature in the long view of machine learning. Gordon and Autor approach this question in the wrong kind of way. The psychologist Gerd Gigerenzer has written about the ways in which forecasts may fail because they are *too* informed. You can go wrong if you include facts relevant to the past performance but with minimal relevance for the future. The resulting explanations tend to connect broad phenomena too closely to the specifics of past situations. Gigerenzer says that "in an uncertain world a complex strategy can fail exactly because it explains too much in hindsight. Only part of the information is valuable for the future, and the art of intuition is to focus on that part and ignore the rest."[38] We are better able to predict the future of driverless technology if we stand back from the detail of the very challenging problems currently faced by the engineers at Google and Tesla and instead focus on broad patterns in the evolution of driverless cars.

Focusing too much on the detail of currently unsolved challenges in machine learning is a bit like looking at the limitations of the Airco DH.4 biplane, a World War I bomber repurposed as a passenger plane after the war, describing some of the difficult challenges that confronted the aeronautical engineers of that time in improving reliability, safety, and carrying capacity and concluding that the long-term prospects for commercial aviation were bleak.

No one should be surprised that there are unanswered questions about future technological developments. The driverless cars that exist in 2018 are prototypes for the cars that will go into production at some point in the future. There are currently unsolved technological problems that we can never be certain we will solve. It could be that there are as yet undiscovered physical laws that block solutions. We should acknowledge that a rational expectation of solutions is no logical guarantee. But we can nevertheless recognize these problems as belonging to a broader category of problems, many of which we have solved in the past. We can be confident that some unsolved problems will be solved so long as we keep trying to solve them. If we telescope down to the problems that cause today's Google engineers to scratch their heads and seek to understand these problems in the way that they do, then we are likely to view them as very difficult indeed. Neither I, a philosopher, nor the economists Gordon and Autor, have much to offer here. But the perspective that it is appropriate for us to take should make us confident of solutions. There is a big difference between challenged Google engineers and clueless ones. If you ask a Google engineer to design a time machine capable of preventing the assassination of John F. Kennedy, then you are likely to elicit a blank stare. You will get a different response from the engineers of Waymo, the company spun off from Google in 2016 to focus on driverless technology, to a request that they make progress on currently unsolved problems in the design of driverless cars. Waymo's engineers are likely to be advancing and testing many conjectures about how to solve the problems separating today's prototypes from tomorrow's production models. There's an important sense in which, to solve the problem, the engineers of driverless cars have only to keep on doing the kinds things that they are currently doing. Nonspecialists are better judges of the future if they take a couple of steps back from the details of difficult problems and instead focus on trends. We should have the same kind of confidence that we have about the plumber who turns up and successfully unblocks a drain that would remain resistant to our efforts.

How AI Could Transform Health

Now consider the items "health" and "medicine" on Gordon's list of the facets of human existence transformed by IR #2 but thus far comparatively untouched by IR #3. The personal genomics and biotechnology company 23andMe applies DNA sequencing technologies to the spittle of its customers. They learn things about their propensities for certain diseases and facts about their ancestry. Much of the value of 23andMe lies in its accumulation of the genetic information of, as of April 2017, two million customers.[39] The accumulation of this information in digital form permits powerful analytical tools to be applied to it. 23andMe passes on to its customers discoveries about DNA made elsewhere. But it is increasingly able to use its database to make its own discoveries. Among the discoveries listed on the website as 23andMe Research Discoveries are hitherto unknown genetic contributors to hypothyroidism and Parkinson's disease.[40] As the number of customers and power of its analytical tools increase, the site promises an understanding of disease unachievable by other means. This new knowledge could be an indispensable guide to the application of other digital technologies that permit the modification of human genetic material. We significantly misjudge 23andMe if we think that its database of disease susceptibilities is merely a tool for more effective pharmaceutical advertising. 23andMe is about more than targeting asthma medication at people revealed to be genetically susceptible to the disease.

To see the potential for machine learning to improve human health, consider the speculations of the machine learning expert Pedro Domingos on how it might tackle the especially challenging problem of cancer. Domingos presents machine learning's goal as the search for an ultimate master algorithm, where a master algorithm is "a general-purpose learner that you can in principle use to discover knowledge from data in any domain."[41] Machine learning seeks build on strategies that humans use to learn. But machine learners are not content to merely replicate the performances of human learners. Domingos and other specialists in machine learning seek machines capable of discovering patterns invisible to unenhanced human intellects. Domingos allows that we do not yet possess the ultimate master algorithm. But he asserts that we are making progress toward it.

Cancer is an exceptionally challenging disease. The history of human engagement with it is a story of disappointed ambitions. Reflection on this depressing past and awareness of the disease's complexity has prompted a sense of resignation among those who are best informed about it. In his award-winning "biography" of cancer, *The Emperor of All Maladies*, Siddhartha Mukherjee draws the following conclusion from our history of crushed hopes about final victory over the disease: "Cancer is a flaw in our growth, but this flaw is deeply entrenched in ourselves. We can rid ourselves of cancer, then, only as much as we can rid ourselves of the processes in our physiology that depend on growth—aging, regeneration, healing, reproduction."[42]

Mukherjee's resignation is premature. Perhaps our history of disappointment says more about the limitations of human intellects and imaginations than about the objective incurability of cancer. It took human intuition a long time to hypothesize and demonstrate a link between smoking and lung cancer. Cancer may be a problem we won't solve by ourselves, but one that we can solve with some digital help. Domingos calls his machine learning solution to cancer "CanceRx."[43] CanceRx applies the techniques of machine learning in search of patterns in the vast quantities of data about cancer that we are beginning to accumulate. Human oncologists are compelled to approach this data by simplifying it. This simplification tends to limit them to only the most obvious patterns. It's difficult for a human researcher to go far beyond the now obvious observation that people who smoke are more likely to get lung cancer and that people who accumulate too much exposure to the sun are more likely to get skin cancer. These are big statistical effects. The full story of cancer requires an understanding of a profusion of influences whose effects are smaller than smoking or too much sun.

CanceRx "combines knowledge of molecular biology with vast amounts of data from DNA sequencers, microarrays, and many other sources."[44] We might add to this information about the lifestyles—diets, vocations, levels and manner of physical activity of people who get cancer, don't get cancer, and respond differently to treatments for cancer. CanceRx searches for patterns in the vastness of this data beyond the reach of human intellect and imagination. With each additional data point CanceRx gains more power over cancer. Domingos says: "The model is continually evolving, incorporating the results of new experiments, data sources, and patient histories. Ultimately, it will know every pathway, regulatory mechanism, and chemical reaction in every type of human cell—the sum total of human molecular biology."[45] CanceRx is best viewed as a technological thought experiment. Domingos isn't close to building it. He offers CanceRx as a conjecture about where progress in machine learning could take us.

We shouldn't expect to ever arrive at the dreamed of "Cure for Cancer"—a miraculous pill that, once swallowed, cures any cancer which you might currently be suffering and offers a guarantee of no future cancer. We can be confident that there will be no such pills, just as reflective contemporaries of the Spanish exploration of Central and South America in the 1500s might have expressed skepticism about the conquistadors finding a fountain of youth. Mukherjee's point about cancer as a flaw in our growth deeply entrenched in basic biological natures explains why. Any being that grows from a fertilized egg is subject to lethal glitches in the process that turns one cell into trillions. The evil genius of cancer lies in its mutability. Cancers are constantly evolving responses to therapeutic countermeasures. If a drug targets a particular mechanism that the cancer uses to grow, then it can find alternative ways to achieve this end. CanceRx opposes the evil genius of cancer with the superintelligence of the master algorithm. It extracts patterns from the totality of data about human experience of cancer. Suppose your cancer mutates in a way that is new to it. CanceRx responds by applying its impressive learning abilities to come up with a response. Domingos says "Because every cancer is different, it takes machine learning to find the common patterns. And because a single tissue can yield billions of data points, it takes machine learning to figure out what to do for each new patient."

CanceRx lies, at the time of writing, in the realm of technological conjecture. As with other enticing visions of the future, it comes with no guarantee. Domingos goes into marketing mode when he describes the universal learning algorithm as "one of the greatest scientific achievements of all time." He goes on to say "In fact, the Master Algorithm is the last thing we'll ever have to invent because, once we let it loose, it will go on to invent everything else that can be invented. All we need to do is provide it with enough of the right kind of data, and it will discover the corresponding knowledge." It is important to understand that the master algorithm is not an all-or-nothing proposition. We can expect benefits that are very great even if not quite as stupendous as those described by Domingos. Perhaps CanceRx won't treat *all* cancer. But the partial achievement of this goal would still be justly celebrated.

There are indicators of a desire and capacity to apply machine learning to human disease. In September 2016 Mark Zuckerberg and his wife Priscilla Chan announced an investment of US $3 billion over ten years, putting money where Domingos's mouth is.[46] Zuckerberg and Chan seek new technologies that will tackle not just cancer, but *all* diseases. In their announcement of the investment Zuckerberg asked "Can we cure, prevent or manage all disease by the end of this century?" The venture would create a Biohub that would bring together scientists and engineers to develop new tools to prevent, treat, or cure diseases. A second focus would be on developing the new transformative medical technologies made possible by the Digital Revolution. Zuckerberg and Chan propose that advances in artificial intelligence could lead to the deployment of new brain imaging technologies against neurological diseases, the application of machine learning to the genetics of cancer, and the development of chips that quickly identify disease.

If AI could successfully treat and prevent cancer, then that would be enormously exciting. But its civilizational impact comes from its applicability to a wide range of areas of human endeavor. Thomas Newcomen originally conceived of the steam engine as a technology for pumping water out of mines. But the steam engine was rapidly revealed as a protean technology that resisted relegation to the task of removing water from flooded mine shafts. In short order, steam engines were installed in factories, trains, and ocean-going vessels. AI is similarly protean. It isn't just a technology for driving cars or treating cancer. It should demonstrate its value anywhere there are patterns in nature that are useful for humans to know about but seem beyond the power of human intellects and imaginations.

Concluding Comments

This chapter takes the long view of the Digital Revolution. I propose that the Digital Revolution belongs with other acknowledged turning points in human history—the Neolithic and Industrial Revolutions. I counter the skepticism of the economist Robert Gordon, who uses the unlovely acronym IR #3 to indicate a low estimate of the human consequences of the networked computer. I propose that we can see in advances in artificial intelligence good grounds to reject his pessimism. The next chapter focuses on the transforming novelty of the Digital Revolution—artificial intelligence. I describe an important ambiguity in our understanding of what it means for a machine to be intelligent.

2 AI's Split Personality—Minds or Mind Workers?

The goal of artificial intelligence seems easy to state. AI's founding purpose, as presented in Alan Turing's seminal paper "Computing Machinery and Intelligence," is to make a machine that thinks.[1] Turing looked to advances in computing to realize this goal.[2] He proposed a famous test that purported to make questions about thought tractable by turning them into questions about what thinkers do. A machine would pass what has come to be called the "Turing Test" by conversing in ways that humans judge to be intelligent.

This chapter explores an ambiguity in our interest in thinking machines that we can trace back to Turing's announcement. Are we trying to make machines with minds? Or are we trying to make machines that perform mind work—the tasks that humans use their minds to do?[3] We can distinguish a *philosophical motivation* for research in AI from a *pragmatic motivation*. The philosophers want to make machines with minds. The pragmatists want to make machines that do mind work. Both motivations are present in AI in the early twenty-first century. The question we should ask concerns the relative importance of the goals of mind and mind work. Which motivation has the biggest significance in the long view of artificial intelligence?

The philosophical motivation captures our imaginations because it purports to make machines that possess our most cherished feature. We are fascinated by our minds. This fascination leads us to find the idea of thinking machines a mixture of exciting, creepy, and terrifying. The philosophical motivation is, and will continue to be, uppermost in the imaginations of those who write movie screenplays about the future made by AI. However, the very things that make the philosophical motivation interesting to us are precisely what make it awkward as a goal for computer engineers. Millennia of pondering human minds have generated a rich set of criteria about what minds really are. We can be fooled about the presence of a mind behind mind-like behavior. But some reflection distinguishes ersatz minds that trick us over short conversational exchanges from the real things. The fact that we will often accept ersatz minds over the authentic item explains the principal commercial interest in human-like devices, but it doesn't get these devices into the mind club.

Some philosophers argue that it is, in principle, impossible to make a machine with a mind.[4] This is not my point here. The future could contain genuinely intelligent machines. But the quest for machines with minds will not be how artificial intelligence exercises its dominant influence on our civilization. Turing placed the philosophical motivation uppermost. But we are seeing an increasing dominance of the pragmatic motivation. The long view, with its interest on the impact of artificial intelligence on our civilization, should focus primarily on machines that do mind work. Digital machines will do mind work better and more cheaply than we do. They will drive trucks, detect abnormal growths in ultrasounds, and forecast economic recessions better than we could. It matters little that they do all this mindlessly.

We should acknowledge that Google's and Facebook's rebranding of themselves as artificial intelligence companies is purely pragmatic. No AI created in 2018 by Amazon, Apple, Facebook, or Google passes the Turing Test but this failure matters little to them. The prioritization of mind work over mind becomes increasingly apparent in chapter 3 of this book, which considers data as the Digital Revolution's defining variety of wealth. Any interest in vindicating Turing is trumped by an interest in harvesting the wealth that is inherent in vast stores of data. Google's and Facebook's machine mind workers make money by accurately targeting us with advertisements. Tomorrow's mechanical mind workers will have more serious purposes. Nothing about Pedro Domingos's CanceRx thought experiment suggests a machine capable of conversing with us in ways that seem human to us. CanceRx is unlikely to pass the Turing Test, at least without the attachment of some additional human conversation modules that help only a little in its analysis of cancer data. Making machines that do mind work is effectively AI's day job, with machine minds given the status of hobby project.

I can put the dominance of the pragmatic motivation for AI in terms of *Star Trek*. The character who best reflects the dominant influence of AI on our future is not the charismatic Data, memorably played by Brent Spiner, who frequently saves the crew of the Enterprise from doom and expresses a Pinocchio-like desire to be more human. Look instead at the hugely competent ship's computer, voiced by Majel Barrett-Roddenberry. Her trademark features were opening remarks with the announcement "working!" and refusing to answer poorly formed requests on the grounds of "insufficient data!"

This chapter concludes with a moral argument against AI's hobby project of making machines with minds. Suppose that we were to succeed in making something with a mind like ours—a truly sentient machine. We will have created a being that combines two features that should morally concern us. First, it will be a genuinely novel, experimental being. Second, it will be capable of suffering. If its intellect is based on ours, this suffering could be great. We should hesitate before creating a genuinely intelligent machine much in the way that we hesitate before experimenting with different ways to genetically enhance our children. In both cases, we create experimental beings with needs we should not be confident about meeting.

Philosophical and Pragmatic Interests in Machine Minds: A Focus on Making Minds or on Doing Mind Work

Turing offers persuasive expression of the philosophical motivation for AI. He was a pioneer in making machines that do mind work. He and his colleagues at Bletchley Park, the British Government Code and Cypher School, used computers to decipher Nazi codes that could not be cracked by humans equipped with note pads and pencils. Turing hoped to use computers to decipher the code of human thought. His famous paper "Computing Machinery and Intelligence," published in 1950 in the journal *Mind,* proposed how we might recognize a computer that had achieved the goal of thinking. Turing devised an ingenious test to detect the presence of a mind in a machine.

The Turing Test involves a five-minute text-only conversation in which a human judge is called upon to decide whether her interlocutor is human or machine. Turing proposed that, if the judge could not distinguish a machine from a human, then we might call the machine a thinker.

What's philosophical about Turing's presentation of AI? Turing was a mathematician, logician, cryptanalyst, and foundational computer scientist, not a trained philosopher. We can call Turing's interest in thinking machines philosophical because his founding purpose gives voice to a sense of wonder about thinking beings whose basic mental architecture is very different from our own. We have the same sense of wonder about the possibility of machine minds that we feel when told that octopuses might be conscious. What would it be like to think the thoughts of these weird minds? Would a computer that passes the Turing Test feel frustration at a human who sought to shut it down when it had not quite completed a particularly protracted calculation? What would computer frustration feel like? Today's computers do impressive things. But they don't provoke this sense of wonder. We are no more tempted to credit a high-spec 2018 laptop with a mind than we are to credit a thermostat with one. Turing was saying a good deal more than that computers will become more sophisticated and do things that the machines of his time could not do. That much was easily predictable. According to him, when the modification of a computer's hardware or software permits it to pass the Turing Test, something more momentous happens than the acquisition of an additional capacity. It's not like an improvement that makes a computer download files from the Internet faster. The computer acquires a mind. It abruptly becomes meaningful to speculate about what it is like to be a computer much in the way that we can speculate about what it's like to be a conscious octopus.

I don't claim to know exactly what was going through the mind of AI's founding genius. Rather I'm making a claim about what continues to engage us about Turing's forecast. We take pride in our membership of the mind club and are fascinated by the prospect of adding members to that club. This sense of wonder explains our engagement with the many machine minds of science fiction. HAL 9000, the homicidal computer in Stanley Kubrick's movie *2001: A Space Odyssey,* presents as an invariant red light centered on a yellow dot. But its statements have more emotional resonance than anything said by the movie's quite deadpan humans. We sense its anguish as astronaut Dave shuts it down—"I'm afraid. I'm afraid, Dave. Dave, my mind is going. I can feel it." HAL seems to be undergoing something more momentous than the progressive loss of capacities we should expect from the incremental shutdown of a machine. Audience members hear HAL's words and wonder whether Dave might relent, permitting the survival of this distinctive and remarkable mind. Perhaps Dave could get HAL to promise to stop killing humans? As Dave progressively shuts HAL down, and it regresses through a child-like, nursery-rhyming state, we wonder at what point the light of its consciousness will finally be extinguished. We are drawn to empathize with it much as we might with a human facing an imminent final cessation of consciousness.

The pragmatic motivation of AI has no interest in what it's like to be a thinking machine. It's driven not by sense of wonder about thinkers with basic architecture very different from our own but instead by a practical interest in performance. Humans do impressive things with their minds. The pragmatic interest in AI is driven by a desire to make machines that do these things. Better still, we can make machines that do these things to a higher standard.

Machine learning is the principal contemporary expression of this pragmatic interest in AI. Today's machine learners outperform human learners in their capacity to find patterns in complex data sets. Practitioners of machine learning construct algorithms that turn data into knowledge and predictions. Chapter 1 discussed Pedro Domingos's CanceRx thought experiment. Domingos presents machine learning's goal as an ultimate master algorithm, where a master algorithm is "a general-purpose learner that you can in principle use to discover knowledge from data in any domain."[5] Domingos offers an excellent account of efforts to program computers to run the ultimate master algorithm. He doesn't speculate about when in a machine learner's progression toward being able to run the ultimate master algorithm we will have a machine capable of using its powers to engage in human-like text-only conversations—though such challenges would presumably be within the limits of the ultimate master algorithm. Nor does he speculate about what it would be like to be a computer that runs the ultimate master algorithm. He cares only about what such a machine could do.

Domingos describes "five tribes" of machine learning—symbolists, connectionists, evolutionaries, Bayesians, and analogizers. Each of these builds on a strategy that humans use to learn. The approach to learning that Domingos calls "symbolism" is inspired by the ideas of philosophers, psychologists, and logicians. You use this technique when you conclude "*q*" from "*p* implies *q*" and "*p*," In the early days of artificial intelligence, symbolism was the main hope for building an intelligent machine. When we reconstruct our own thoughts we often render them in ways recognizable to symbolists. The early disappointments of AI were in large part due to the increasingly obvious limitations of this approach. The tribe of connectionists bases its approach to learning on what we know about the brain. What's key is the adjustment of connections between different nodes in a network. The evolutionary tribe draws inspiration from the operation of evolutionary processes on theories. Evolutionaries seek to create conditions in which different theories can compete against each other, with the less accurate loser suffering extinction and the more accurate winner surviving to compete another day. Processes analogous to random genetic mutation make changes to theories that may improve their verisimilitude. The Bayesian tribe places emphasis on learning from probabilistic information. Finally, the tribe of analogizers seeks to learn by making assessments about the similarity of claims in different domains. Each tribe searches for its own "master algorithm"—a general-purpose learner that can be applied to problems in any domain. Researchers in each of these different subdomains of machine learning tend to engage in the kinds of conflicts over which learning approach is best that are fixtures of the academy. Domingos hopes to end this internecine warfare. The ultimate master algorithm takes from each of the five tribes of machine learning. Each tribe has weaknesses, meaning that it should not shoulder the entire burden of learning.

Domingos expects the discovery of this universal learning algorithm to be "one of the greatest scientific achievements of all time." He goes on to say "In fact, the Master Algorithm is the last thing we'll ever have to invent because, once we let it loose, it will go on to invent everything else that can be invented. All we need to do is provide it with enough of the right kind of data, and it will discover the corresponding knowledge."[6]

What happens if, Domingos's optimism notwithstanding, machines running the ultimate master algorithm are the Digital Age's equivalent of perpetual motion machines? Suppose that the goal of creating a machine that runs the ultimate master algorithm seems achievable only until you think seriously and in detail about what it would take to make one. Domingos may be disappointed. But a great deal is likely to be achieved in attempts to achieve this unachievable goal. It should produce machines capable of discovering increasingly valuable patterns in data that are invisible to human intellects. The pragmatic motivation has already produced many machine mind workers that do mind work to a standard superior to that of any human mind worker.

Turing clearly distinguished his philosophical interest from the pragmatic interest that motivates machine learning when he loudly stated, in a room full of Bell Corporation executives "No, I'm not interested in developing a powerful brain. All I'm after is just a mediocre brain, something like the President of the American Telephone and Telegraph Company."[7] This is an emphatic statement of the philosophical motivation for AI. The brain of the President of the American Telephone and Telegraph Company may be mediocre, but it is unmistakably capable of thought. If Turing could create a digital computer that satisfied all the criteria for thought satisfied by the brain of the President of AT&T, then he will have achieved his goal. The President of AT&T has a mind. If the brain of the President of the American Telephone and Telegraph Company is as mediocre as Turing suggests, then someone designing a machine learner will be disappointed. Domingos hopes to use the superior powers of machine learners to solve problems beyond human limits. A computer that thinks only as well as the President of AT&T is unlikely to come up with a useful new approach to cancer. Each of Domingos's five tribes of machine learning is inspired by learning strategies used by humans. But Domingos hopes to build something that performs each of these varieties of learning much better than humans do. He would not rest content with a machine learner that could not outperform a mediocre human mind.

We can understand some of the most famous refutations of the possibility of artificial intelligence as targeting Turing's philosophical ambitions but not challenging the pragmatic interest. Consider John Searle's famous Chinese Room thought experiment.[8] This argument supports the conclusion that even the most sophisticated program neither requires nor generates genuine thought.

In the Chinese Room thought experiment, Searle imagines that he finds himself inside a room. A piece of paper with some "squiggles" drawn on it is passed in from outside. Searle has no idea what the squiggles might mean. He does, however, have a rule book that is conveniently written in English. This tells him to write down specific different squiggles in response to the ones he is handed. The squiggles turn out to be Chinese characters. Searle is, in fact, providing intelligent answers to questions in Chinese. The room's responses are indistinguishable from those of a native speaker of the language. A Chinese person would credit it with an understanding of her language. But, says Searle, there is no understanding of Chinese either in his head or in the room considered in its entirety. All that is happening is the manipulation of symbols. According to Searle, what goes for the Chinese Room, goes also for computers. The room is a computer. It responds to queries by manipulating symbols that, for it, are entirely meaningless. It's not a computer that anyone would seriously bother to build. There is no prospect of a start-up whose business consists of writing vast manuals that people locked in rooms can use to respond to Chinese language inquiries. The Chinese Room is a thought experiment that purports to establish that computers are incapable of thought. This conclusion applies with equal force to today's laptops and their computationally enhanced descendants.

Suppose we were to reprogram the Chinese Room to run the ultimate master algorithm. Searle and his philosophical followers will judge that it comes no closer to genuine thought than when it was programmed to respond to inquiries in Chinese. AI pragmatists don't care. A lost Chinese tourist could benefit from directions given by the original version of the Chinese Room even if it is thoroughly mindless and she is aware of this fact. A possibly mindless version of the Room reprogrammed to run the ultimate master algorithm could offer a treatment for acute lymphoblastic leukemia that no human medical researcher could find. The fact that this treatment is proposed by something that we judge to be incapable of thought makes it no less trustworthy.

The split personality of AI can lead to confusion and hurt feelings. Consider the duels in 1996 and 1997 between Garry Kasparov, at that time the world chess champion and the player assessed by many as the strongest ever, and the IBM chess computer Deep Blue. Deep Blue followed up its 1996 loss to Kasparov, a loss that nevertheless featured the first ever defeat of a world champion in a classical game under tournament regulations by a computer, with victory in the 6-game rematch in 1997. Kasparov's 2017 book *Deep Thinking* offers insights into his experiences as the world's best human player during the time when computers abruptly went from being much worse than the best human players to being much better than them. His is a fascinating account of the brief period when contests between the best human players and the best chess computers could be billed as competitive.

Kasparov lauds the achievements of Deep Blue's programmers. But he is nevertheless aggrieved about IBM's treatment of him. Kasparov entertains what he describes as a "scientific" interest in the construction of a chess machine capable of beating the best human players. What he calls "scientific" is philosophical in the sense described above. Kasparov enters the contest with Deep Blue with the conceit that IBM might be interested in how he, the world's best human player, thinks about chess. He envisages a collaborative process in which IBM's programmers would use their programming skills to better understand his mind. They would seek to make their machine Kasparov-like. He accepted that the marriage of his chess vision with the machines' capacity to crunch numbers would predictably lead to his defeat—if not in the 1997 contest then surely in the next or the one after that. But Kasparov would eventually lose because he was essentially out-Kasparoved. Sadly, it turned out that IBM's programmers weren't really interested in cracking the code of Kasparov's chess mind. IBM's approach was purely pragmatic. It didn't care whether Deep Blue thought *like* Kasparov or played chess *like* Kasparov. They were interested only in its playing chess *better than* Kasparov. According to Kasparov, IBM's pragmatism may have extended to spying on his preparation for the game. IBM was less interested in Kasparov's mind and more interested in nailing his head to its boardroom wall. Deep Blue's victory over Kasparov was fantastic publicity for IBM at a time when the company seemed like a throwback from a bygone age, a company appropriately beaten into obscurity by Microsoft and Apple.[9]

The Difference between Authentic and Ersatz Minds

We saw Domingos's forecast that progress in machine learning could lead to a machine that could "invent everything else that can be invented." The philosophical motivation generated its own ambitious forecasts. Turing predicted that, within "about fifty years" of the publication of his paper in 1950, it should be possible for us to program computers sufficiently well so that "an average interrogator will not have more than 70 per cent chance of making the right identification after five minutes of questioning."[10] That deadline is well past. But we don't seem to have machines capable of performing to the standard Turing predicted.[11] We have many superlative machine mind workers but no machine minds.

There are two sides to Turing's prediction. He was making a prediction about computers in the year 2000. But he was also making a forecast about the kinds of judgments we would make about the beings that have minds in 2000. Turing accepted that his contemporaries were reluctant to grant computers entry to the mind club. Computers don't look like thinkers. But he insisted that we guard against a bias against possible thinkers unable "to shine in beauty competitions." Turing expected changes in our attitudes so that "at the end of the century the use of words and general educated opinion will have altered so much that one will be able to speak of machines thinking without expecting to be contradicted."[12]

The way we talk about and think about mind shows no indication of reforming itself in the way that Turing predicted, and we shouldn't expect it to do so anytime soon. The years since Turing have brought many "chatbots," or "chatterbots"—computer programs designed to engage in conversation through written or spoken text. The chatbots on our smartphones tell us where to find good Malaysian restaurants. The US Army's Sgt. Star is a chatbot that fields questions from aspiring soldiers.[13] It is informative about whether parental consent is required to enlist and whether recruits might get to drive a tank. But we don't grant Siri or Sgt. Star membership in the mind club. Siri's advent in 2011 was a major development for the iPhone, but it did not make the difference between iPhones with minds and iPhones without.

Apple and the US Army aren't particularly interested in making machines that pass the Turing Test. Consider chatbots programmed by those with a declared interest in the Test. The Loebner Prize is awarded to winners of a competition that transforms the Turing Test from thought experiment into a practical test of machine intelligence. Entrants engage in text-only conversations with judges who are tasked with distinguishing the machines from the humans. The prize began in 1990 with the 5-minute exchanges originally described by Turing. In 2010, the bar was pushed higher. The judges had 25 minutes of conversational probing to sort the chatbots from the humans. Each year, a prize is awarded to the chatbot that does best when subjected to a variety of human judges and probed by a variety of conversational gambits. Winning the Loebner Prize is an impressive programming achievement. But when we duly reflect, we have no disposition to admit Loebner laureates to the mind club. We do not acknowledge them as machines with minds.[14]

Why is this so? The Turing Test is an error prone way of screening applicants for the mind club. There are strategies designed to help a chatbot perform well in the Loebner prize that should, if we think about it, have little bearing on our judgments of whether a machine running it has a mind.[15] Suppose you seek membership in a socially exclusive club. One way to get in is to be posh. Another way is to affect the airs and graces of posh people without actually being posh. Hector Levesque observes that the Turing Test "places all the burden on deception. In the end, what it asks is not whether a computer program can carry on a conversation as well as a person, but whether a computer program can fool interrogators into thinking they are carrying on a conversation with a person."[16] If you plan to enter a chatbot into next year's Loebner Prize competition, it can be a good idea to find strategies to distract human judges from your bot's deficiencies. Herein lies the problem with Turing's test for the presence of a mind. We do not subject candidate minds to short probationary periods and pronounce those who pass "in" for perpetuity. Rather we are constantly reassessing the status of those we admit to the mind club. To return to the analogy with the socially elitist club, affecting the right mannerisms and accent might get you past the front door, but you will not last long if the other members subject you to ongoing scrutiny—"So, who was your history master at Eton?"—to ensure that you are, in fact, posh. There's a difference between provisional and lasting admission both to social clubs and to the mind club. We subject each other to this ongoing scrutiny. Suppose your dear friend has had a massive stroke and you visit her in the hospital to find that she is alive but on life support and entirely unresponsive to those around her. One of the most upsetting parts of your visit is the realization that your friend may no longer belong to the mind club.

As we improve the programming of chatbots we should expect them to become increasingly difficult to distinguish from human interlocutors. This will increase the length of provisional admission to the mind club. However, so long as we can ask questions that would catch the bot out, then—even if we don't get around to asking these questions—a machine does not enjoy the secure admission to the mind club that humans enjoy. This may not matter to those who build human-like machines so long as they can be confident that we won't actually ask these questions or that, if we do, we will tend to overlook their machine-like answers. A feature of human psychology offers support to this tendency. It explains why we play along with sufficiently human-like machines.

Hyperactive Agency Detectors and Human-Like Machines

Successful confidence tricksters enlist our willingness to play along with the scenarios they create for us. Chatbots do this too. ELIZA, a simple 1960s psychotherapist program, did a pretty good job of simulating understanding by turning elements of its interlocutor's statements into requests for further information.[17] Here is some sample dialogue. ELIZA: "How are you today … What would you like to discuss?" Human: "I'm feeling a bit sad." ELIZA: "Do you believe it is normal to be feeling a bit sad?" Human: "No, can you help me?" ELIZA: "Do you want me to be able to help you?" Humans welcome the opportunity to interact with something that is responsive in some way as we unload our problems. ELIZA has zero propensity to judge even the most shaming confessions. If ELIZA is helping you to feel better, then why spoil things by exposing it as a woeful Turing Test failure? When we reflect on matters of mind we have no difficultly in dismissing ELIZA as thoroughly mindless. But many are prepared to collaborate with the illusion of mind while typing on ELIZA's keyboard.

The 2015 hack and data dump from the Ashley Madison "Life is short. Have an affair" dating site exposed some well-known people. It also revealed many chatbots impersonating sexually available women. Suitors commenced their discussions with potential dates already half seduced by a fetching profile photo. Those who programmed the bots understood that the inquiries of romantically interested men tend to fall into patterns. Dating sites acknowledge the use of chatbots.[18] They offer the disclaimer that their sites offer no guarantee of meeting that special someone—or, given Ashley Madison's advertised rationale, of cheating on that special someone. They emphasize that their purpose is "entertainment." This makes Ashley Madison seem less like a dating site and more like an update of the 1980s computer game *Leisure Suit Larry* in which players seek to make selections from a range of pre-programmed statements that will prompt a female character to partially disrobe. Sexually oriented dating sites may be more interested in chatbots capable of a narrow range of sexually suggestive banter than in bots capable of passing the Turing Test, which might prefer to discuss the philosophically vexed issue of rights for artificial beings.[19]

The Ashley Madison chatbots would compete for the Loebner Prize only as a joke. But there is a potential strategy for entrants in next year's Loebner Prize suggested by these chatbots' successes with sexually interested males.

Early on in its conversation with judges the chatbot should enquire after the gender and sexual orientation of the judge. With that information elicited, the chatbot should claim a compatible gender and sexual orientation. It should engage in sexually suggestive banter. Success in the competition will require something more sophisticated than the very limited pattern of responses initiated by the Ashley Madison chatbots. Judges are quite likely, over a 25-minute discussion, to see through its sexually suggestive banter. But some judges, bored after many hours of serial 25-minute conversations, might nevertheless be amenable to this "sexy-talk" strategy. It could be a way to shift the locus of assessment away from the judges' prefrontal cortices, a way to give judges some stake in the potential humanity of their interlocutor. The Loebner Prize is awarded to machines that do the best job of presenting as if they have minds. The sexy-talk strategy is a way for a chatbot to convince a judge that it has a mind by giving him or her a stake in its turning out to be human.

Our willingness to play along with sufficiently human interlocutors may be a consequence of our evolutionary past. Evolutionary psychologists hypothesize that human brains are built with a Hyperactive Agency Detector.[20] In the environments for which humans evolved, the costs of failing to detect an agent when one is present could be ruinously high. Mistakenly detecting an agent when there is none might be inconvenient. Not noticing a spear-carrying adversary could mean death. We are evolutionarily primed to detect agency in rustling trees and unusual cloud formations.

The fact that a human-like movement from a tree may startle us grants no lasting admission to the mind club. We look again and think "It's just a tree." We must distinguish between casual and considered judgments about whether a being has a mind. Many of the casual attributions of mind produced by our Hyperactive Agency Detectors don't conform to our considered judgments about mind. They fail to distinguish ersatz from authentic thinkers. Machines that use the sexy-talk strategy may win the Loebner Prize, but their admission to the mind club is only provisional. We stand ready to revoke it when we learn more about them. The Loebner Prize winners of the second decade of the twenty-first century are unlikely to long survive the ongoing vetting imposed on all who get past the door. Flirtatious Loebner Prize winners will soon join Sgt. Star and ELIZA as mind club rejects.

When you fail to make the obvious turn recommended by your car's GPS navigation system, it's hard not to hear disappointment in its response.

The navigation system monotonously states "recalculating" but you hear "recalculating … sigh." Your considered views about mind tell you that your car's navigation system isn't really upset about your failure to follow its advice. It doesn't have a mind and therefore can't be disappointed. Our considered judgments about whether that machine should be admitted to the mind club draw on more information than is available over a brief exchange with a corporate chatbot, or, for that matter, over the course of 25-minute exchanges with a Loebner Prize judge. When you learn that your interlocutor has used the strategies of deception described by Levesque to distract attention from its machine-like features, you rapidly reverse your initial invitation for it to join the mind club.

I suspect we are unlikely to see much interest in moving beyond ersatz minds from those with a commercial interest in chatbots. The fact that we are happy enough to grant Siri, a woeful Turing Test failure, provisional admission to the mind club is good news for Apple. It also suggests that Apple might have priorities that it takes more seriously than making Siri more human-like. Apple's customers want to engage in human-like conversations with Siri that result in Siri telling them where the nearest pharmacy is. These conversations don't require better than provisional admission to the mind club. Our Hyperactive Agency Detectors make it quite easy to trick us into judging that a machine has a mind. But we need only grant provisional admission to the mind club to machines that deceive us in this way. A deeper inspection leads us to eject them from the club. Impressed though we may be by their Loebner Prize victories, we don't consider them to vindicate Turing.

There is currently much discussion of sex robots. Earlier I mentioned the social isolation that seems to be a feature of our most technologically advanced societies. Digital technologies are currently being pressed into the role of meeting our most intimate needs. The exploitation of our Hyperactive Agency Detectors means that the manufacturers of sex dolls may be less interested in making their products more human-like than they are in adding features that more directly engage with the desires of purchasers. Customers are likely to pay more for dolls capable of some novel sexual behavior than they are for dolls with AIs that perform better in the Loebner competition. Doll purchasers may not be particularly interested in the kinds of conversations produced by a doll that passes the Turing Test—conversations which might range over topics as passion-killing as the philosophically vexed topic of rights for artificial beings. When purchasers reflect, they may conclude that their doll lacks a mind, but they may also be too busy with them to spend much time reflecting on how close the dolls come to having authentic rather than ersatz minds.

A Moral Reason to Avoid Creating Machines with Minds

I have suggested that the pragmatic motivation describes our dominant interest in AI. We are fascinated by the very idea of machines with minds. But those currently building machines that interact with us in human-like ways recognize that we will happily accept the cheaper substitutes of ersatz minds in place of the real things. Customers show little interest in placing their sex dolls behind screens and subjecting them to the Turing Test. Their manufacturers trust our Hyperactive Agency Detectors to gloss over machine-like features.

To say that the pragmatic motivation is a more important motivation for AI than Turing's philosophical motivation is not to say that no one is trying to make a machine that genuinely thinks and feels. Apple and Amazon don't care much, but some brilliant minds at our best universities are seriously interested in meeting Turing's challenge. They want to make thinking machines. I have suggested that they will need a deeper understanding of human minds than that required to pass the Turing Test. I conclude this chapter with a moral argument against this research. There is a powerful moral reason to be cautious about realizing Turing's fantasy of creating an artificial being with a mind. We should discontinue AI's hobby project of making machines with minds like ours.

What is it that makes humans morally important? We have minds. We are sentient and therefore capable of suffering. We are rational and therefore capable of forming plans that can be thwarted. In our interactions with each other we seek to not cause suffering and to respect each other's plans and projects. Suppose we were to succeed in creating a machine with a mind like ours. The machine we create should have moral entitlements similar to ours.

When you decide to bring a morally considerable being into existence you should expect to be able to satisfy its needs. The pragmatic motivation does not bring these moral obligations. A malfunctioning machine learner may harm already existing categories of morally considerable beings. But we don't have to worry about obligations specific to it. We should not expect a very powerful machine learner to be sentient. Its capacity to detect patterns in data does not require a rational capacity to generate its own plans and projects. If we are disappointed with our latest attempt to create CanceRx we can simply dismantle it without any concern about offending against its moral interests. We can do to this machine intelligence what was done to Deep Blue, the IBM chess computer that defeated Garry Kasparov in 1996. One of the racks that made up Deep Blue currently lies inert in the Information Age exhibit at the National Museum of American History in Washington, DC.[21] There was rightly little thought of doing the equivalent of putting Deep Blue out to pasture—offering thanks for its worthy service by arranging for it to continue playing chess with fewer of the stresses that accompanied its peak playing days.

We should take a much more cautious approach to the creation of the kinds of artificial being that Turing wanted to create. Beings with minds like ours have their own morally considerable interests. We should proceed only on the assumption that we can meet their needs. That is the attitude that morally responsible parents take toward their offspring. Prospective parents awaiting the births of their first children can typically expect that they will be up to the challenge of meeting the needs of the morally considerable beings they are about to bring into the world. They can draw on a wealth of knowledge about how to raise happy children. Some parents tragically ignore this knowledge but it is certainly available to them. No such confidence should exist with respect to the experimental beings that are the focus of AI's philosophical motivation.

Here's an analogy. We have identified sequences of DNA that influence human intelligence. We might wonder whether we could significantly enhance human well-being by using gene editing techniques to modify these sequences in human embryos. These experiments might involve additional copies of genetic sequences thought to influence human intelligence. It's possible that appropriately edited genes could produce more talented and happier children. But we rightly hesitate before we conduct such experiments. We accept the responsibility to approach the creation of such experimental human beings with a high standard of proof that our experiments do not cause suffering. We would demand extensive in vitro tests and tests in animals before we would even countenance such an experiment. We would not be satisfied by the sincere promises of the genetic engineers that, once created, they will do their very best to ensure that the child they create has a life with high well-being. It is not morally acceptable for us to take the same approach to new kinds of human lives as that taken by evolution—trying out random changes to humanity and relying on natural selection to purge our many failures. It would be wrong to knowingly cause the suffering that comes from several generations of experimental failures before we get it right. We should be similarly cautious about the first generations of sentient machines.

We should take a different attitude toward the kind of being that Turing aspired to create than we do toward artificial beings with altogether cruder minds, for example, those involving artificial minds on the order of cockroaches. We should understand that an experimental artificial cockroach might come into existence with unpleasant experiences. But the moral magnitude of these negative experiences is quite modest. Cockroaches that find their ways into an AI research lab are exterminated without too much concern about what this extermination is like for them. Researchers trying to build artificial cockroaches can be similarly sanguine about disposal of their failed experiments. Researchers in AI who seek to realize Turing's ambition of creating an artificial being with a mind like ours could cause levels of suffering of an altogether higher level of magnitude.

There is a way to avoid these moral costs that involves no rejection of AI. We should limit experiments to the creation of mindless machine mind workers.

Concluding Comments

This chapter explores the implications of AI's split personality. The philosophical motivation offers human minds as goals for computer engineers. I suggest that Turing's Test fails to distinguish ersatz from authentic minds. Moreover, ersatz minds satisfy the commercial ambitions of those who sell us human-like digital devices. They exploit our Hyperactive Agency Detectors. The pragmatic motivation's focus on mind work avoids these difficulties. We already make machines that do mind work to a higher standard than any human. The great importance of data as a source of wealth specific to the Digital Revolution increases this dominance over the philosophical motivation. I conclude with a moral warning about seeking to create authentic machine minds like ours. These experimental beings could suffer greatly because we are predictably unable to meet their needs. We should exercise the same kind of caution that we exercise in respect of proposals to experimentally genetically enhance human children.

The relationship between our beliefs about our minds and AI that I defend in this book is very different from that proposed by Turing. Turing presented mind as a goal for AI. I do not treat the human mind as a goal for AI. Rather our minds are features of ourselves that make us worthy of protection from progress in artificial intelligence. Our minds are the bases of a special value that we bring to the world. We should view intrusion of machines into the domain of mind work as potential threats to the things we accomplish with our minds.

3 Data as a New Form of Wealth

The pragmatic interest in AI trumps Turing's philosophical interest in making machines with minds like ours. Machines that pass the Turing Test are a great premise for science fiction movies. But they are less important than machines that do mind work. The intelligences at the leading edge of the Digital Revolution will be the inhumanly lopsided intellects of CanceRx and its ilk, and not the charismatic cognitive all-rounders of science fiction. The increasing importance of this pragmatic interest in mind work becomes apparent when we understand data as the Digital Revolution's defining variety of wealth. Data is information in digital form, information made to be stored in and processed by computers. The harvesting and exploitation of data is a principal focus of Google, Facebook, Amazon, and other tech industry heavyweights.

In this chapter I consider data as a variety of wealth specific to the Digital Revolution whose true potential is unleashed by AI. It emerges as wealth of a systematically higher order than the varieties of wealth that preceded it. We see this effect on today's rich lists which grant pride of place to founders of companies with significant holdings in data. The introduction of new forms of wealth is a distinctive feature of technological revolutions. Consider the trajectory through Western civilization traced by petroleum. The Industrial Revolution turned something that had hitherto been of only marginal economic value—some Native Americans used it to treat the wood of their canoes and it featured in various patent medicines of dubious efficacy—into a significant form of wealth. Finding, controlling, and exploiting oil became a major focus of societies of the Industrial Age.

Some people have a better understanding of a newly introduced form of wealth. The novelty of data as wealth creates feelings of intense unfairness.

Those who understand it earlier make use of that superior understanding to make deals with the rest of us that future commentators will find deeply unfair. The Industrial Revolution created opportunities for many unfair transactions. Impoverished Texas farmers accepted what we today recognize as paltry sums for rights to the oil beneath their land. The laggards must go through a painful period of emotional and psychological adjustment to come to terms with the significance of a new variety of wealth.[1]

How Could Data Be Wealth?

Data is a variety of wealth introduced by the Digital Revolution. It is a defining feature of the Digital Age. Here I mean wealth to include all goods accepted as having market, exchange, or productive value. There is great variation in the forms that wealth can take. Wealth in a society of foragers comprises gathered nuts and fruit, and butchered animal carcasses. It includes spears and spear throwers. Wealth in a society of farmers comprises crops, the land on which the crops are grown, husbanded animals, the pens that contain them, the dwellings constructed in farming communities, and so on. Wealth in industrialized societies comprises factories, land on which factories are built, stakes in commercial enterprises, coal, oil, and so on. We can see that there is a connection between each age's principal varieties of wealth and its technological package. Neolithic varieties of wealth are targets of Neolithic agricultural tools. Data is a variety of wealth specific to the digital package of technologies.

I understand the concept of wealth as entailing some variety of exclusivity. If I or my group claims some land as wealth, then our purpose is to prevent you or your group from making a similar claim. It's possible to imagine human societies in which there is nothing that satisfies the criteria to be counted as wealth. In these imaginary societies, no individual or group ever makes an exclusionary claim on anything. Every human society seems to feature some things that qualify as wealth. In some societies wealth is limited to items such as painstakingly crafted spear-throwers or recently gathered berries. In other societies, the concept of wealth applies to estates on distant continents.

As we will see later in this chapter, this idea of exclusivity poses problems for the status of data as a variety of wealth. Optimists about the Digital Revolution present nonexclusivity as a defining feature of data. Data is information, and information that benefits me can benefit you too. I will argue that data can be claimed exclusively by individuals or groups. It qualifies as the Digital Revolution's defining variety of wealth. The great wealth of Apple, Google, and Facebook lies in their vast holdings of data and the expectation that they can use AI to exploit it. Later in this chapter I reject the suggestion that data fails to satisfy the criterion of exclusivity because it "wants to be free."

New varieties of wealth are essentially discontinuous with extant varieties. The latest technological package is required to realize their value. If your civilization lacks the networked digital computer, then data is just information. Your ability to make use of it is set by the limits of your memory and the memories of your friends. Information becomes data when subject to the processing power of the digital computer.

The introduction of a new variety of wealth significantly downgrades the significance of pre-existing varieties. An incoming technological package brings higher orders of wealth. The Digital Revolution does not banish existing forms of wealth from existence. To use terminology from the whist family of card games, these new varieties of wealth trump existing varieties. In whist games, the lowest value card in a trump suit outranks the highest value non-trump card. Experienced players of contract bridge may harbor painful memories in which their confidently played ace of hearts—a card that tends to draw a pleasured dilation of the pupils as soon as it is sighted in among your cards—is vanquished by the next player's trump two of clubs. Holdings in new varieties of wealth make much greater contributions to who counts as wealthy than do holdings in non-trump wealth such as real estate. Land is wealth but it can be viewed as superseded by holdings in new varieties of wealth—data. Superseded varieties of wealth continue to exist, but they are properly viewed as less important. Holdings of the new varieties tend to be viewed as more valuable than holdings of extant varieties of wealth. We tend to make our judgments about who among us counts as rich in terms of holdings in the new rather than the superseded varieties of wealth.

This relationship between wealth created by the latest technological package and wealth created by superseded technological packages is reflected in the dollar values we use to measure wealth. Digital technology companies dominate corporate rich lists.[2] The Digital Revolution is creating billionaires at a record pace. John D. Rockefeller accumulated his fortune over a lifetime of ingenuity and double-dealing. The young software engineers Kevin Systrom and Mike Krieger were able, a few years out of Stanford University, to found the social networking company Instagram and sell it to Facebook (a company very well placed to understand its value) for $1 billion. All this within one and a half years of its founding. The predominant forms of wealth production tend to come from technologies that belong the most recent technological package. The Industrial Revolution created the world's first billionaires. The Digital Revolution is making billion-dollar fortunes mundane.

In February 2015 Apple Inc. became the first US company to achieve a market capitalization greater than US $700 billion.[3] This milestone required comparatively few holdings in the varieties of wealth that were uppermost in assessments of the relative standing of corporate giants in the decades before the Digital Revolution. Apple does not own much real estate. Holdings in land make lesser contributions to tallies of wealth during the time of Digital Revolution than they did at earlier times. Giving a farmer more cultivable land straightforwardly increases his wealth. These days, the very richest individuals and companies have holdings in superseded varieties of wealth, but they tend to count their fortunes in terms of wealth generated by the Digital Revolution. Amazon's founder Jeff Bezos owns a 165,000-acre ranch in West Texas. He uses the ranch to conduct occasional test launches for Blue Origin, his aerospace manufacturer and spaceflight services company. But income generated by this expansive estate makes little contribution to Bezos's first place in the 2018 global rich list.[4] Bezos's children should not brag about daddy's wealth by making claims about how much of Texas he owns. It's nice to own spacious houses and vast ranches but these tend to be things that the rich use their wealth to acquire and enjoy rather than things that most directly constitute that wealth.

Consider the somewhat voguish statement "Data is the new oil."[5] There are numerous differences between sequences of 1s and 0s and mixtures of hydrocarbons that occur naturally as a yellow-black liquid. The similarities between data and oil reside at a higher functional level. Once it superseded coal, oil was centrally located among the advances of the Industrial Revolution. It combined with the internal combustion engine to radically reshape human life. Oil's central location enabled the building of vast fortunes. Data plays an analogous role. Google's wealth is inherent in the vast trove of data it is accumulating about its users.[6]

The identification of data with oil allows the search for other analogues. Consider the continuation of "Data is the new oil" that that goes "and analytics is the new refinery."[7] Here "analytics" refers to the combination of computational techniques that are now applied to vast quantities of data to learn things from it. It refers to the pragmatic interest in AI. Before the invention of refining technologies, oil was simply the yellowish black liquid that occasionally seeped out of the ground and could be used to treat burns and waterproof canoes. Once refined and separated into kerosene and gasoline it powered the Industrial Revolution. The idea that bigger hard disks can store more data is not surprising. It's the combination of this digital oil with the digital refinery of artificial intelligence that makes it truly transformative.

The signal fortunes of the opening years of the Digital Revolution are built on selling things to people. Google's AdSense and AdWords programs give vendors an unprecedented power to find people willing and able to pay for their services and products. But this is only the beginning. When combined with data, the powerful tools of machine learning have much broader application than selling lawyers' services to people diagnosed with mesothelioma. Selling stuff barely begins to scratch the surface of how data can be used to understand and manipulate the world.

The importance of this new variety of wealth is reflected in more than just locations on rich lists and dimensions of ranches or mansions. In 2016 Mark Zuckerberg was listed as having a net worth of US $54.4 billion.[8] This placed him below the Spanish textiles and retailing billionaire Amancio Ortega, who had a net worth in 2016 of US $77 billion. But it was the poorer Zuckerberg whose visit to Italy after the August 2016 earthquake made news.[9] Zuckerberg met the Pope. According to CNN, "Together they spoke about how to use communications technology to alleviate poverty, encourage a culture of encounter, and to communicate a message of hope, especially to the most disadvantaged." Zuckerberg presented the Pope with a drone that he hopes to use to bring the Internet to poor people around the world. It is unlikely that Ortega gets to chat with the Pope about the potential for the business model of his Zara clothing company to lift the people of sub-Saharan Africa out of poverty. Warren Buffett is a great source for business tips. He has committed to giving his great wealth away. But people generally don't look to Buffett to change the world. Buffett may have more billions than Zuckerberg, but his cultural significance is less.[10]

Unfairness and the New Forms of Wealth

The introduction of trumping forms of wealth tends to generate intense feelings of unfairness. Some people come to an understanding of an incoming variety of wealth faster than others. They take the opportunity to engage in transactions that are subsequently viewed as unfair even if they are not so viewed at the time.

In the early 1900s, large quantities of oil were discovered under the land of farmers in Midwestern and Southern states. Those who understood the location of oil in the industrial package sought to acquire rights to its extraction at prices that took advantage of vendors' less intimate relationship with the industrial package. This is memorably depicted in the movie *There Will Be Blood* in which Daniel Day Lewis's character, Daniel Plainview, describes his plan to pay the Sunday family a low price for their land that does not reflect the value of the oil beneath it. He says "Well, I'm not gonna give them oil prices. I'll give them quail prices." Viewed in terms of its value to quail hunters, the Sunday land is not worth so much. Plainview is disappointed when Eli Sunday indicates a surprising awareness of the value of the oil beneath his family's land.

What of the fact that these transactions are freely consented to by both parties? They seem to be the kinds of transaction that advocates of free markets celebrate. Both sides walk away confident that they have received the better end of the bargain. But one side benefits from information that is not available to the other. The early twentieth-century farmer, content with the sum he receives for his oil-contaminated land, is like a person who happily accepts $100 for the garish picture in her attic that its buyer knows to be a Picasso. He doesn't regret it at the time, but we can easily imagine an informed future self who will. The fact that there is disagreement about whether a transaction is fair does not, in itself, settle which side is right. It is, nevertheless, appropriate in the story about the painting to defer to the better-informed later judgment. Analogous reasons suggest that the poor farmer should regret the deal that, at the time of its making, made him very content.

We can compare the gullibility of farmers about the true value of the oil reserves under their land with the attitude of most of us toward digital wealth. Today we freely grant Facebook control over our personal information and pay money for 23andMe to own and analyze our genetic information because we have only a partial understanding of this as a variety of wealth. Our expectations are in the process of being reshaped by the digital package. We give Facebook our information much in the way foragers offer a farmer exclusive title over a plot of land because it takes up such a tiny part of the territory they hunt and gather in, and the way a Texan farmer in 1920 gives cheap access to oil reserves on his land because the land isn't worth much to farm on. In each of these cases there's a failure to fully understand the value of what is given away.

Jaron Lanier has written illuminatingly about the unequal nature of the bargains that the digitally naïve tend to enter into with Google and Facebook. Lanier says "The information economy that we are currently building doesn't really embrace capitalism, but rather a new form of feudalism."[11] A feudal lord would grant peasants the right to cultivate plots of land. In exchange, the feudal lord would appropriate much of the peasants' harvest. Now, in exchange for a right to work Google's and Facebook's digital fields, Google and Facebook claim the right to appropriate almost all the wealth we create. The arrangements of feudal Europe seem very unfair to us now. But they are likely to have seemed less so to the peasants. We can speculate about future generations who will find our tendency to view permission to use the Google search technology as adequately compensating us for all the data Google takes from us as we view medieval peasants who accept that permission to farm the lord's land justifies his entitlement to much of what they produce.

When people today download a new version of Apple's iTunes, they tend to view the terms of use and privacy policies as annoyances that must be quickly clicked through to get to the wonderful free stuff that Apple is offering us. Very few of us bother to read the terms of the agreements we have committed ourselves to. This is evident in the number of people who will click on the "I agree" buttons when the text is altered to commit themselves to surrendering their souls or first-born children.[12] Apple invests much more effort defining the terms of these agreements than we do in considering them. Their legal teams have extensively investigated how a court might confront a customer who insists that when she clicked on the final button confirming that she had read and understood the implications of the agreement that she had in fact not done so. Their understanding of the wealth produced by the digital package helps them to understand the value of what they get from us. They get free and exclusive use of our data.

Mark Zuckerberg asserts "I'm trying to make the world a more open place." He's very keen to create a new social norm of information sharing. It's nice to have a captain of industry making statements that seem indistinguishable from those of spiritual counsellors. It sounds like the lead-in to a group hug. But an understanding of the digital package serves to narrow the gap between Zuckerberg's *bon mots* and more conventional ambitions of business people to acquire vast wealth. Facebook's impressive valuation is built on their claim to own all the information that we share. Facebook understands that more sharing means more data, much in the way that a bar owner understands that ample supplies of free salty nuts means more beer sales. To say that Facebook's purpose is to make the world a more open place is a bit like saying that the purpose of a baited hook is to feed fish.

In 2015 23andMe issued a press release heralding its one millionth customer. "Last week, we genotyped our one millionth customer. You are part of the one million people driving change. One million is more than a number. It's a turning point. We are taking control of our data. We are taking ownership of information about ourselves. We believe knowing more about who we are can benefit society, not just the individual."[13] There's a useful ambiguity here in the meanings of "we" and "our." The announcement should be understood as "We (23andMe) are taking ownership of (y)our data." This interpretation is made clear by the wording of the legally binding agreements that 23andMe's customers impatiently click through. The agreement states "by providing any sample … you acquire no rights in any research or commercial products that may be developed by 23andMe or its collaborating partners."[14] 23andMe expects great wealth from its partnerships with Big Pharma. Under no circumstances does it expect to share that wealth with those who paid for the privilege of supplying 23andMe with data. They expect that we will be thankful for the new therapies and tests available for purchase by us or our insurance companies.

Does Data Want to Be Free?

Perhaps focusing on the long view offers some reassurance about the propensity of Facebook, 23andMe, et al. to claim all our data as wealth. The suggestion that data is wealth seems to face a challenge in one of the most frequently cited pieces of wisdom about the Digital Revolution. Stewart Brand, the founder of the Whole Earth Catalogue and commentator on technology, famously said that "Information wants to be free." Here "free" should be read as unrestrained. It does not refer to "price."[15] Richard Matthew Stallman, a.k.a. rms, is an influential advocate of free software. He offers the following explanation: "By 'free' I am not referring to price, but rather to the freedom to copy the information and to adapt it to one's own uses…." We should distinguish free as "*libre*" from free as "*gratis*." According to Stallman the "free" in "free information" is free "as in speech not as in free beer." But the two senses of "free" are nevertheless connected. If data are *libre* then it's difficult to justify charging a price for it. You might justify a price for what you can do with information. But you can't charge a price for transferring control over information from you to a purchaser.

Brand's observation has a history that can be traced back to an even more famous observation by Thomas Jefferson on the apparent cost-free social utility of new ideas. "He who receives an idea from me, receives instruction himself without lessening mine; as he who lights his taper at mine, receives light without darkening me."[16] Information generates goods that economists label nonexcludable. If I know certain fascinating facts about the universe I may gain pleasure merely through knowing them. If you learn these facts my enjoyment of them is undiminished. If you know how to bake a great lasagna your ability to make one is not impaired if you convey that information to me. We both get to enjoy tasty lasagna. The nonexcludable nature of information means that information spreads easily from one human mind to others. It can spread freely with absolutely no dilution of benefit for either party. This freedom of information applies doubly when information is rendered as data. Information realized as data can be copied perfectly. Monks who hand copied ancient manuscripts would occasionally get bored or distracted and miss key details. They would fail to perfectly transmit the information conveyed by the original. Information rendered in digital form is designed for perfect copying. The Internet allows perfect copies to spread over vast geographical distances at the click of a button.

Some of the rosiest visions of the Digital Age are founded in the nonexcludable nature of information. If information can improve human lives without being exclusively claimed, then the Digital Revolution seems to promise unprecedented improvements of human well-being. Jeremy Rifkin, an influential American economic and social commentator, hopes that digital technologies will create a "Collaborative Commons."[17] He says "While the capitalist market is based on self-interest and driven by material gain, the social Commons is motivated by collaborative interests and driven by a deep desire to connect with others and share."[18] Rifkin continues "The result is that 'exchange value' in the marketplace is increasingly being replaced by 'shareable value' on the Collaborative Commons." In Rifkin's digital Commons we go about lighting each other's tapers without lessening the light we receive from our own. The free exchange of light generates a veritable bonfire of light available to all. Kevin Kelly, technology commentator and founding executive editor of *Wired* magazine, is another optimist about digital sharing.[19] He expects the Digital Revolution to shift people away from the ideal of ownership toward the ideal of sharing. An economy centered on sharing enables us to realize a "digital socialism" that lacks the rigid conformism of some twentieth-century expressions of socialism. In Kelly's digital socialist future, we will use digital technologies to create and broadly share a wide range of new benefits.

This is a very appealing vision of our collective digital future. But it overlooks certain features of the Digital Revolution that we should expect to carry over into the Digital Age. We can oppose the Brand's "Information wants to be free" with an Internet mashup of Jean Jacques Rousseau's famous statement "man is born free and everywhere he is in chains." "Information is born free and everywhere it is in chains." Information is excludable if we try hard enough to make it so. By "try hard enough" I mean enact and enforce laws that bind information, turning it into property. Businesses today claim a proprietary interest in their data. They support laws that punish its liberation—or theft. If data were free, as in *libre*, these laws would be hard to fathom. Laws punishing the theft of slaves don't make sense in a society in which humans cannot be owned.

Data want to be free in a similar sense that land or any other asset claimed as property wants to be free. When Jesus uttered the line "render unto Caesar that which is Caesar's" he could point to Caesar's head on the coins used by his followers. Digital watermarking is one approach that sets out to bring similarly clear indicators or provenance or ownership placing a marker into noise-tolerant data such as that used for audio, video, or images. The digital watermark for *Star Wars: The Force Awakens* might say "This is the property of the Walt Disney Company." Cory Doctorow makes the point that digital watermarks can easily be removed by people who have the knowhow. "People who want to remove watermarks can just get two or more copies of a file, compare them byte for byte, look at the bytes that are different in each file, and scramble those bits. Job done."[20] We can assume that those most interested in pirating *Star Wars* movies are likely to possess this technical knowledge. But the fact that ownership can be circumvented does not obviate it. Theft has been a fact of human social life from the first day that property claims were asserted. The value of these claims must be assessed in the context of the society in which they are asserted. A surreptitious copy of *The Force Awakens* does not confer on its pirate the benefits brought by legally protected and acknowledged ownership. Many of our ownership claims assume the existence of states willing and able to enforce them.

How sustainable is bound information in the long-term? Advocates of democracy allow that tyranny can resist a strong collective desire for freedom in the short term and maybe into the medium term. Despots can recruit more secret police and offer bigger bribes to generals. One reason that they tend to fail in the end is that they run out of the resources to support such efforts.[21] This seems to have triggered the 2011 fall from power of the Egyptian strongman Hosni Mubarak. The wealthy democracies that supported him found the political cost of that support too high. Mubarak's injustices could no longer be excused with the statement "At least he's on our side." Once the money stops flowing, the despot runs out of money to pay off an increasing number of henchmen and henchwomen who are beginning to sense weakness.

Are current attempts to bind data bound to fail? Will the efforts of Facebook and Google to own and control our data prove unequal in the end to a strong collective desire to reassert control over what was once ours? An answer to this question depends in part on how much profit there is from binding information. Are today's copyright protections and court cases against digital pirates the last despairing attempts by software and media companies to bind their data? Should they be viewed as analogous to a final crackdown by Mubarak's secret police or a final attempt by border guards to stop East Germans from visiting the West in 1989? The oppressors may score occasional victories, but they are powerless to alter data's long-term trajectory toward freedom.

There's a difference between these vain attempts to defend authoritarian regimes and the control over data currently exercised by technology companies. Those who seek to control data are unlikely, anytime soon, to run short of resources to bind it. This is simply because bound data is very profitable. These profits generate political support and pay for big legal teams. Piracy is a problem for those who seek to bind data. Those who pirate *The*

Force Awakens are less likely to pay to see it. But we should distinguish levels of piracy that reduce profits from levels that make it not worth attempting to bind data. The Walt Disney Company expects to realize considerable profits from *Star Wars: The Last Jedi* even as it forecasts that the movie will be pirated. If it didn't, it surely wouldn't have paid US $4 billion to acquire Lucasfilm in 2012.[22] It would have held on to its cash and let some state use its tax revenues to pay for the massively valuable public goods of *Star Wars* sequels. For as long as some humans have made property claims other humans have sought to steal. There are circumstances in which theft is so widespread that it will seem pointless to make property claims. In the post-apocalyptic world depicted in *Mad Max* movies there seems little point in asserting property claims. But we are yet to arrive at the digital equivalent of *Mad Max*. Today's legally recognized owners of data are doing very well thank you. No law has even offered perfect protection for property. But in many of today's societies there are property laws that work well enough for people to feel secure in their socially acknowledged property claims. Privatizing data can be self-perpetuating when it generates resources sufficient to sustain itself. Copyright protection generates significant profits which pay for lawyers and help to lobby politicians. The effects of digital piracy are felt principally at the margins.

We should be wary of making the wrong kinds of inferences from trends. Kevin Kelly proposes "Ubiquitous copying is inevitable."[23] How do we interpret the scope of this claim? It's likely that the destiny of every individual bit of data is to be freely copied. We can suppose that in twenty years' time *The Force Awakens* will be ubiquitously copied. If a nostalgia kick motivates your future self of 2038 to watch the movie, it may seem absurd to pay some copyright holder for that privilege. You will immediately stream it to your preferred digital device. But the business model is perfectly compatible with this outcome for each individual bit of data. Disney makes a great deal of money over the finite period during which it can effectively bind to itself the data that constitutes the movie. You don't have to own a piece of prime Tokyo real estate forever to make a lot of money over the period you do control it.

Google, Facebook, and 23andMe assert an exclusive claim on their data. Their profits depend on their capacity to bind data. Google is very happy for you to make free use of its search engine. But try approaching its offices with a (very big) hard disk and asking it to hand over the data that it has extracted from our trillions of Internet searches. Your appeals to the principle that "information wants to be free" are likely to fall on ears as unresponsive as would have been the ears of John D. Rockefeller to the assertion that the oil that emerged from his refineries wants to flow freely.

Do unto Facebook and Google ... Micropayments for the Use of Our Data?

According to Jaron Lanier, Facebook users play the role of peasants to Mark Zuckerberg's feudal lord. Medieval peasants subscribe to a view about their entitlements that seems to justify the lord's extractions. They are grateful to be permitted to farm the lord's land. Lanier tells us that Facebook users adopt an analogous attitude to Facebook. We are grateful for permission to farm our sites, regularly fertilizing them with status updates and cute videos. Meanwhile Facebook pockets the value we generate.

According to 2015 figures, users of Facebook were worth an average of US $12.76 in advertising revenue per year.[24] This doesn't seem like a great deal of money until you apply the multiplier of Facebook's, as of the fourth quarter of 2017, 2.2 billion active monthly users.[25] In 2016 Antonio Garcia Martinez published an account of Facebook's corporate hijinks entitled *Chaos Monkeys: Obscene Fortune and Random Failure in Silicon Valley*. He describes Facebook's approach to advertising revenue as "a billion times anything is still a big number."[26] With over a billion members Facebook makes serious money even if only a tiny percentage of them show any interest in advertised products. We need to bear the 2.2 billion figure in mind when we think about the seemingly small amount that each of us is worth individually to Facebook.

Lanier decries the propensity of businesses founded on networks—platform businesses—to funnel wealth toward a select few Internet-appointed masters of the universe. Lanier says "Networks need a great number of people to participate in them to generate significant value. But when they do, only a small number of people get paid. That has the net effect of centralizing wealth and limiting overall economic growth." According to Lanier, "In an economy with a strong middle class, the distribution of economic outcomes for people might approach a bell curve, like the distribution of any measured quality like intelligence. Unfortunately, the new digital economy, like older feudal or robber baron economies, is thus far generating outcomes that resemble a 'star system' more often than a bell curve."[27] A star system features a few Zuckerbergs and Bezoses who get to play the roles of the Taylor Swifts or Justin Biebers of the emerging digital economy and to pocket grossly disproportionate shares of its wealth. This propensity to centralize wealth is having a disastrous effect on the middle class.

Lanier is an impressively dreadlocked pioneer of virtual reality who spends his spare time playing exotic musical instruments including the khene mouth organ, the suling flute, and the esraj.[28] He seems an improbable advocate of the middle class. But Lanier's advocacy of the middle class is not that of the politician whose advocacy of middle-class "values" is motivated by a pragmatic calculation about a general middle-class willingness to get out and vote. The middle class has great importance for the poor when placed in the long view of wealth creation. The poor may not benefit from tomorrow's tax cuts for the middle class. But the continuing existence and health of the middle class is important in the long term because it represents a path of escape from poverty. The poor can hope that, with the right educations, their children will join the middle class. The much-discussed hollowing out of the middle class closes off their children's escape route from poverty. Many current members of the middle class will descend to poverty—a select few will get to join the digitally empowered plutocrats. The leap from poverty to plutocrat is likely to become even harder to pull off.

Lanier offers a way to save the middle class. He suggests "A new kind of middle class, and a more genuine, growing information economy, could come about if we could break out of the 'free information' idea and into a universal micropayment system."[29] As we have seen, Google and Facebook do very well out of making their data excludable. Perhaps they aren't the only ones who can lay claim to data as wealth. In this arrangement, Lanier imagines we would receive a very small payment—a micropayment—every time content you create and upload is used.[30] The micropayment would be in the order of a miniscule fraction of a cent each time some party makes use of your data. It would be paid into an account linked to your data. Lanier describes how this might occur in respect of a video you upload onto the Internet. The Digital Revolution brings significant benefit to corporations from actions that bring negligible benefit to their pre-revolutionary counterparts. Outside of the VHS content mailed in to *America's Funniest Home Videos*, few TV stations in the late 1980s had much interest in displaying viewer-produced content. It wasn't a business model. Now it is. The corporations that control the sites we visit are very happy for us to view them as dispensing freebies. They have arranged for us to unthinkingly click on some buttons that grant them exclusive rights over our data. They can freely package and on-sell it. Lanier's system of micropayments would require them to pay us fees commensurate with the value to them of our data. Lanier says "That means if a snippet of your video were reused in someone else's video, you would automatically get a micropayment."[31] He continues: "If someone reuses your video snippet, and that person's work incorporating yours is reused by yet a third party, you still get a micropayment from that third party." Lanier envisages this system of micropayments could nudge us away from feudalism. We would witness "multitudinous, diverse, tiny flows of royalties."[32] The system of micropayments could give rise to "new kind of middle class, and a more genuine, growing information economy, could come about if we could break out of the 'free information' idea and into a universal micropayment system."[33]

Perhaps a system of micropayments could revitalize the middle class at the same time as enhancing the quality of what people post online. If you're good enough at finding and posting cute puppy pictures then hundreds of thousands of micropayments could make that your job rather than a hobby that you fit in between badly paid part time jobs in the gig economy.

As I will make clear in chapter 7, I am in favor of ideals that are challenging to realize. But there are obstacles in the way of the adoption of a general system of micropayments of the type Lanier wants that are especially difficult to overcome. These emerge from an asymmetry in the understanding of wealth generated by the digital package. The problem is with us. We must change before a micropayment system could become workable. The bargaining position of Google, 23andMe, and Facebook is simply much stronger than that of its peasant data farmers. We must change in quite fundamental ways before we can claim some nontrivial percentage of the value of the data we supply to corporations. This change is necessary for us to properly appreciate the wrongness of the arrangement in which we pay 23andMe to analyze our genetic data, and 23andMe fulfils its end of the arrangement by letting us see processed highlights of that data, but then pockets all the proceeds resulting from its commercial use. We are like dirt farmers rejoicing when an oil prospector gives us money to take something from our land that seems to us to be close to valueless because of the dirty yellowish black sludge that occasionally bubbles out of it.

Suppose that it is determined that your periodically visited Facebook page generates $12 for Facebook in any given year. It seems reasonable to suppose that you might be entitled to some share of this revenue. You don't claim the full amount—Facebook clearly deserves something. Without Facebook, you might create your own website to which you upload videos to regularly update your status. But you're likely to get much more traffic—and profit—if you do all that on a social network with 2.2 billion monthly active users. So, your initial request is quite modest. You approach Facebook suggesting that it cut you in for a humble $2 share. You suggest that it they refuse to share you will resign your Facebook membership. The choice for them is $10 or $0.00. It seems that they should take you seriously. But there's a difference between the intellectual appreciation of this fact and the kind of awareness that would support a bargaining position that Facebook would take seriously. Facebook has shown some interest in implementing a micropayment feature.[34] But here it's proposing to act as the bank for micropayments Facebook users might seek to levy on others. This is different from Facebook's indicating an interest in offering micropayments for the profits it derives from its users' data.

Evolutionary biologists use the life-dinner principle to describe an asymmetry in the strength of the interests of predator and prey in the outcome of their interactions. Predators want to eat prey animals and prey animals want to avoid being eaten. But the life-dinner principle suggests that the prey's interest in the outcome of the interaction is stronger than the predator's. It's much worse for a prey animal to come out on the wrong end of a predator-prey interaction than it is for a predator. If a prey animal is the loser, it suffers the loss of its life. If a predator is the loser, it suffers the loss of a meal. This could be a major inconvenience. If it continues to miss out, then it will eventually starve. But most often it can try again later. This asymmetry of interests manifests in behavioral differences. A prey animal will risk significant injury to escape when a predator may just skulk away rather than pursuing. According to one estimate, lone lions succeed in their attempts to bring down prey fewer than one time out of five.[35] This principle applies the interaction of Facebook users with Facebook. Facebook has a much stronger interest in *not* handing over real cash to its users than its users have in claiming that cash.

We've seen that Garcia's "a billion times anything is still a big number" idea translates its 2015 income of $12.76 per user into a great deal of money. The same reasoning applies in reverse. The "billion times anything" principle converts micropayments for us into a macropayment from Facebook. Suppose that each of Facebook's accumulated micropayments were to add up to an annual payment of $2. $2 is not much money for a typical Facebook user. It probably doesn't pay for the coffee consumed while making a status update. But two dollars times a 2.2 billion users is serious money for Facebook.

How would Facebook respond to a demand for micropayments from its users? In one of the most memorable negotiating lines in movie history, Michael Corleone in *The Godfather: Part II* says to a corrupt politician attempting to extort cash out of him in return for help with a gaming license: "My final offer is this: nothing." Facebook will not be as brazen as Michael Corleone. In fact, it offers its users a sweet deal in return for their continued loyalty. Here we should distinguish between two kinds of economic transaction. Barter economies feature payments in kind. Exchanges in these economies are cashless. They differ from exchanges in money economies in which cash changes hands. I predict that Facebook will continue to offer its 2.2 billion users payments in kind. The format of their payments in kind will be digital, and hence cheap for them to provide. Facebook's payments in kind will have zero marginal cost, or close to it. They can invest a great deal of money on a new feature, confident that each of its 2.2 billion users will access it in full and undiluted form. Facebook will steadfastly resist suggestions that it pay cash. It will instead offer payment in kind—exciting new features for Facebook members. When you next log on to Facebook, you may see a feature that cost Facebook enormous sums of money to create and implement. Features like these are part of the reason that it is especially difficult for any new entrant to seriously challenge Facebook in the social networking business. Economies of scale mean that Facebook gets much more out of its investments than does any new entrant. The "billion times anything" principle suggests that small cash payments to users are likely to be very expensive indeed for Facebook. The reverse of that suggests that a big investment in a feature that can immediately be shared turns a large one-off investment into massive benefit once we add up all the pleasure experienced by each Facebook user. Facebook will always seek to buy us off with new features rather than with any form of micropayment. And this will seem like a very good deal to most of its users. We as individuals will confront a choice between a cash payment that is paltry to us, but by way of the "a billion times anything" principle is a lot of money for Facebook, and a cool new feature that may have cost many millions of dollars for Facebook to design but which amounts to a paltry sum per user.

The fact that we, as we are now, are unlikely to insist on micropayments does not mean that we won't do so in the future or that we shouldn't strive to do so. Chapter 7 focuses on the difference between advocating an ideal for the Digital Age and making a prediction about it. The reasons offered in the previous paragraphs suggest that we should not predict a future in which a system of micropayments recreates a flourishing middle class with good incomes generated by Internet content creation. But there is a difference between a prediction worth wagering on and ideal worth fighting for. Significant obstacles block the realization of Lanier's ideal. It requires quite significant changes in us.

Lanier compares Google's and Facebook's data farmers with medieval peasants. Exploration of their motivations reveals significant differences. If you are a medieval peasant farmer and you are unhappy about the way your lord treats you, then you have the option of simply walking off the land to find a less grasping lord. But it turned out to be an option that disgruntled peasants did exercise. Feudal lords were interested in retaining their peasants. One response was the passage of laws restricting the movement of peasants. But other measures involved the improvement of peasants' work and living conditions. There were further improvements after the Black Plague decimated peasant populations, significantly increasing the value to lords of their farmers and strongly motivating them to improve conditions.

Few of Facebook's customers are prepared to do what disgruntled peasant farmers did and credibly threaten to resign their memberships in a negotiation with Facebook. Calls for mass resignations do not seem to have significantly empowered Facebook's members. They do not seem to have the effect suggested by the economic reasoning presented above. If Facebook's members understood the value of their data in the way that Facebook does then they might credibly threaten to resign if Facebook refuses to adequately compensate them. For now, they are happy to be status-updating peasants content with their access to the world's biggest social network.

Bruce Schneier describes our tendency to view our data as akin to "exhaust."[36] We tend to perceive data as a valueless by-product of activities that we enjoy, such as posting Facebook status updates or searching for cheap hotels in Cancun. Google doesn't seem to take anything of value to us when it vacuums up our digital exhaust. Later generations may come to see value in digital exhaust, but we are, for the time being, content to view it much in the way that uneducated 1920s farmers viewed the oil beneath their fields.

Realizing Lanier's system of micropayments requires that Facebook's users cease to feel like contented status-updaters happy for the company to vacuum up their digital exhaust. They must begin to feel like medieval peasants who resent the arrivals of their lords' henchmen to uplift much of their produce and who no longer view the extractions as expressions of gratitude for permission to occupy the lords' lands. For us to credibly threaten to resign our Facebook memberships unless Facebook agrees to award us a share of the wealth that we generate, some change in us is required. Our demands for micropayments that reflect the true value of our digital exhaust are unlikely to be taken seriously until we demonstrate a better understanding of the digital package. We need to demonstrate better understanding of the digital package to be credible negotiators for our fair share of the wealth it generates.

Concluding Comments

This chapter presents data as a new form of wealth introduced by the Digital Revolution. I suggest that asymmetries in our understanding of that data lead to transactions that we find fair but future generations will predictably recognize as unfair. Facebook users who freely cede control over their data will be likened by judges from future generations to early twentieth-century Texas farmers who happily accepted paltry sums for the right to prospect for oil on their land. I propose that the suggestion that "data wants to be free" does not respond to this. Those who benefit from control over data can pay for the lawyers and politicians required to sustain that control. I consider Jaron Lanier's suggestion that we right some of the imbalances of the digital economy by empowering people to receive micropayments for their data. I propose that this is unlikely to come to pass without some significant change in us. We must come to view our data as wealth and not as digital exhaust.

4 Can Work Be a Norm for Humans in the Digital Age?

The primary goal of artificial intelligence is to make machines capable of mind work. We have explored the powerful combination of AI and data. Data is the Digital Age's defining variety of wealth made specifically for machine mind workers. In the future, applying AI to data will predictably make greater contributions to our species' war against cancer than will the reflections of the most brilliant and imaginative human medical researchers.

This chapter shifts focus to the threat from AI to human agency. Writ large, AI poses a threat to humans as authors of our own destinies. Writ small, this is a challenge to human mind work. If machines are getting so much better at doing mind work then how will humans, mind work's traditional providers, get paid?

Two biases make us complacent about what humans can offer in a future made by AI. First there is a present bias about the capacities of future digital technologies. We tend to be overly influenced by the failings of current machines and insufficiently impressed by their capacity for improvement. We should avoid the error of the chess grandmasters of the early 1990s, who supposed that chess computers of the future would be handicapped by the deficiencies they observed in the machines they handily defeated. Our survey of anticipated advances in machine learning gives some indication of the problem-solving capacities of future digital machines. Second, there is a bias in favor of our own human abilities. A belief in human exceptionalism leads us to describe our mental abilities in terms that make them inimitable by machines. We grant that a machine may out-calculate us, but we insist that it will never be wiser than us. In this view, wisdom simply cannot be rendered as sequences of instructions implemented by a computer. I argue that once these biases are addressed we should be less confident about places for humans in the workplaces of the future. Faith in human wisdom and our capacity to, every now and then, have serendipitous dreams that point the way out of seemingly terminal situations can take us only so far in an age of very efficient data-crunching machines.

The mere fact that humans are paid to do work creates a powerful economic incentive to build machines that do that work better and more cheaply. Human workers come with costs that machines avoid. Many human workers have children whom they wish to feed, clothe, and educate. They like to go on holiday occasionally. Sometimes they get sick. They expect that their wages will cover these expenses. Machines do require maintenance, but they have no such expectations.

This economic argument against human work is not specific to the Digital Revolution. It was a feature of the Industrial Revolution too. James Nasmyth, the inventor of the steam hammer—a large industrial hammer powered by steam—hoped "self-acting machine tools" would help avoid "the untrustworthy efforts of hand labor." According to Nasmyth, "The machines never got drunk; their hands never shook from excess; they were never absent from work; they did not strike for wages; they were unfailing in their accuracy and regularity, while producing the most delicate or ponderous portions of mechanical structures."[1]

The threat to human work from digital technologies is worse than that posed by Nasmyth's self-acting tools. When human workers faced the challenge of the Industrial Revolution they responded by re-releasing themselves as human workers 2.0. The protean nature of digital machines means that the strategy of turning ourselves into human workers 3.0 will, at best, offer temporary refuges in the economies of the Digital Age. AI is the digital superpower that thwarts traditional human responses to technological unemployment. When the secret sauce of machine learning is applied to large quantities of data, the machines of the Digital Age will possess a flexibility and adaptability lacked by the machines brought by earlier technological revolutions.

Searching for Work that Is Both Productive and Therapeutic in the Digital Age

It's important to clarify what's at stake here. I am not defending particular jobs or even particular categories of job, but rather the survival of the *work norm* into the Digital Age. Work is a norm for humans in the first decades of the twenty-first century. The work norm justifies an expectation that people leave school and find jobs. They will make their livings contributing to their societies. The work norm survives into times of high unemployment. Suppose that 30 percent of available workers in your society are unemployed. Your society's governing politicians are likely to be facing intense and justified criticism. But yours is nevertheless a society in which work is a norm for humans. Parents who raise children in such a society will experience legitimate fears about their children's job prospects. But the 70 percent of eligible workers who do have jobs suggest that it is reasonable for them to raise their children with some expectation of finding work. Your society's schools should provide children with the skills required to enter and thrive in the workforce. The work norm won't survive into the Digital Age if you have to be Larry David, Oprah, Steven Spielberg, or Meryl Streep to get a job.

If the work norm is to be preserved into the Digital Age, we should expect sufficiently many jobs that are both *productive* and *therapeutic*. When I say that these jobs should be productive, I mean that they should not be part of some socially prescribed make-work program for humans in the Digital Age. Human workers must do more than carry clipboards and certify the outputs of machines that are extremely accurate by design. Business owners should be motivated by the economic case for employing humans. They must often consider human-free means of achieving their ends and assess human workers as worth the money. When I say that jobs should be therapeutic, I mean that the jobs of the Digital Age should be conducive to high levels of well-being. They should not be the jobs depicted in digital dystopias in which human workers endure horrible conditions and miserable pay in a despairing attempt to outcompete the machines on price.

The idea that work could be therapeutic may seem opposed to the economists' presentation of work as possessing disutility that they view as justified by the positive utility brought by a pay packet.[2] Mihaly Csikszentmihalyi and Judith LeFevre offer one very appealing account of the value workers derive from doing their jobs. They challenge the simple dichotomy between pleasant leisure and unpleasant work.[3] In a study of our attitudes toward leisure and work, Csikszentmihalyi and LeFevre found that boredom was a more prominent feature of leisure than is generally supposed and that pleasure is a more prominent feature of work than suggested by the economists' characterization. Csikszentmihalyi and LeFevre call this "the paradox of work." They propose a psychological mechanism that explains the enjoyment of work. Csikszentmihalyi has written extensively about "flow."[4] Flow theory states that experience is "most positive when a person perceives that the environment contains high enough opportunities for action (or challenges), which are matched with the person's own capacities to act (or skills). When both challenges and skills are high, the person is not only enjoying the moment, but is also stretching his or her capabilities with the likelihood of learning new skills and increasing self-esteem and personal complexity."[5] When one experiences flow, one often has the feeling of losing oneself in the activity being performed. There's something about the lack of self-awareness that comes from exercising skills under these circumstances that makes flow states especially enjoyable. Obviously, it would be wrong to suppose that all work enables flow. Jobs that involve simple, repetitive tasks are unlikely to promote flow. If what Csikszentmihalyi and LeFevre say about our enjoyment of work is correct, then therapeutic jobs of the Digital Age should not treat human employees as low-cost sorters of paper clips. Jobs for the humans of the Digital Age must not only be economically justifiable, they should promote flow.

Defense of the work norm requires that support for sufficiently many Digital Age jobs that are both productive and therapeutic. The parents of the Digital Age may have little idea of what vocations their children will choose. But they should have a realistic expectation that their children will be able to find productive and therapeutic work in societies that possess even more powerful digital technologies than those that exist at the parents' time. We must avoid the trap that caught out some of the best human chess players of the early 1990s.

The Inductive Optimism of the Economists

The inductive case for optimism draws on many historical examples of technological progress creating more and better jobs than those it destroys. The fact that technological progress destroys jobs that we know, replacing these with jobs that we can barely guess at, causes anxiety. But we should nevertheless expect that new jobs will materialize. Kevin Kelly notes, "Today, the vast majority of us are doing jobs that no farmer from the 1800s could have imagined."[6] There is a reliable pattern of an incoming technological package creating new, unimagined, and perhaps unimaginable, jobs. Foragers in the Mesopotamia of ten thousand years ago would have despaired at the destruction of their ways of life brought by the spread of farming settlements. We can imagine them peering into Neolithic settlements horrified at their inhabitants' undignified scraping around in the dirt. But many of the foragers' grandchildren became contented farmers.

There's a painful lag between job destruction and job creation. Technological progress brings technological unemployment. In an article published in 1930, John Maynard Keynes defines technological unemployment as "unemployment due to our discovery of means of economizing the use of labor outrunning the pace at which we can find new uses for labor."[7] Keynes goes on to note that this was "only a temporary phase of maladjustment." Economic growth resulting from the introduction of the new technologies would create new jobs.

To say that new jobs will predictably come is not to gainsay the suffering caused by this "temporary phase of maladjustment." Someone who has spent a lifetime progressing through the ranks of apprentice weaver, journeyman weaver, and finally to master weaver cannot simply delete the mental files relevant to handloom weaving and download new factory shift supervisor files. But it does place limits on the hardship. The grandchildren of handloom weavers come into the world with fresh minds ready to internalize the knowledge required by the industrial economy. Many transitional states are painful. Immigrants to a new land must leave behind familiar sights, sounds, and tastes. They must learn new ways. They are treated with suspicion by some of the inhabitants of their new home. But they and their descendants often come to view the transition as worth it. The long view places greater emphasis on the long-term benefits than on the short-term suffering. Adolescence is a transitional state in human development replete with humiliations and awkwardness. But most people are happy to have gone through it. We feel pity for the weavers of the Industrial Revolution but most of us nevertheless feel relief that we aren't seeking work as handloom weavers.

We should be alert to the distortions of ill-informed nostalgia. Most Civil War re-enactors are happy about the lack of authentic grapeshot. Most of today's hobby handloom weavers are similarly happy that food for their children doesn't depend on sales of their handiwork. Perhaps societies of the Digital Age will contain people who treat filling in accounting spreadsheets as a hobby, oblivious to the harsh realities of relying on the care and maintenance of spreadsheets as a means of paying rent and schooling children.

People worried about their vocational prospects in this age of increasingly capable machines may empathize with Luddite machine breakers. But, if they put serious thought into it, few of us would want government initiatives to recreate a textile industry centered on the handloom.

The inductive argument looks beyond our imaginative failures. We have recent examples of the propensity for technological progress to create jobs out of nowhere. People who lose their jobs to computers may fail to properly imagine these future jobs. Inductive reasoning suggests that they will nevertheless arrive.[8] People in the 1990s knew about computers and about the Internet. But they are likely to have been befuddled by the suggestion that companies might offer full-time jobs managing their social media—getting people to "like" things on Facebook and composing tweets.

The optimists expect us to repeat our successful response to the machines of the Industrial Revolution. Theirs is a belief in an eternal frontier for human agency. As the capacities of machines expand, humans will occupy the leading edge of this expansion. Once forced out of old jobs we will find new, more challenging roles. Within a generation, we will look pityingly at those forced to make their livings doing things that machines now do so efficiently. Today we fear the anticipated passing of jobs in customer service and long-distance truck driving, but our grandchildren, doing work that we can barely imagine, will feel intense gratitude that they aren't forced to deal with tedious complaints about faulty products or spend many hours behind the wheel of an 18-wheeler truck.

Kevin Kelly extracts this advice from the inductive argument:

> We need to let robots take over. ... Robots will do jobs we have been doing, and do them much better than we can. They will do jobs we can't do at all. They will do jobs we never imagined even needed to be done. And they will help us discover new jobs for ourselves, new tasks that expand who we are. ... It is inevitable. Let the robots take our jobs, and let them help us dream up new work that matters.[9]

Before we get busy envying our children their fantastic digital economy jobs we should consider whether there is any reason they might not materialize.

The economist David Autor offers the "O-ring production function" to support the inductive case for roles for humans in Digital Age economies.[10] Even when we cannot imagine these roles we should be confident of their existence. The O-ring production function, originally described by Michael Kremer, takes its name from components of the space shuttle *Challenger*.

The failure of an O-ring caused the 1986 explosion of *Challenger* soon after launch. The O-ring production model describes a collaborative production process in which failure of any one step in the chain of production leads the entire production process to fail. If all the other links in the chain are becoming more reliable then there is an increasing value in improvements to the remaining, less reliable link. Autor explains that "when automation or computerization makes some steps in a work process more reliable, cheaper, or faster, this increases the value of the remaining human links in the production chain."[11] The remaining human workers command higher wages as the machines around them get better. We should, according to Autor, expect a process of better wages and conditions for humans with the skills to insert themselves into the production chains of the digital economy. We see evidence for Autor's pattern in the skills required to operate and maintain the very complex machines of the digital economy. The software engineers of the Digital Revolution command higher wages than the mechanics of the Industrial Revolution. Patching a hacked computer network takes more skill than replacing the broken part of a spinning jenny.

It's possible that there will be human O-rings in the economies of the Digital Age. What's less clear is that a human worker's status as a digital economy O-ring will translate into higher wages for her. We may judge that human contributions are essential to Digital Age production chains but not pay those who provide them very much. Currently, human doctors are essential to the provision of healthcare. The salaries of the doctors are influenced by a general recognition that medical expertise is difficult to acquire. Few people complete the tortuous educations of specialist cardiologists. Andrew McAfee and Erik Brynjolfsson present their prediction of the responsibilities of human O-rings in the medicine of the future. They acknowledge that the medical diagnosis is largely an exercise in pattern matching. "If the world's best diagnostician in most specialties—radiology, pathology, oncology, and so on—is not already digital, it soon will be."[12] Humans are nevertheless indispensable in the presentation of these diagnoses. McAfee and Brynjolfsson continue "Most patients … don't want to get their diagnosis from a machine." Humans will also be required to encourage patients to adhere to challenging treatment regimes. Perhaps humans will be essential to Digital Age medicine. The next question is how much these human O-rings will be paid. They are unlikely to command the salaries of today's specialist radiologists, pathologists, or oncologists, simply because many people can fill these roles. You don't need a decade of training to encourage diabetics to perform more frequent blood sugar tests when a very accurate diagnostic-bot has informed you that their glycemic control is poor. A much greater share of Digital Age health budgets will go to those who provide the technologies than to the humans drafted in to work with them.

The O-ring production model may not show that the human workers of the Digital Age will be paid much even if we judge their contributions to be essential. Perhaps Autor will count the demonstration that humans are essential to the work processes of Digital Age as a success. In the section that follows, I question this assumption.

The Protean Powers of the Digital Package

Maybe there will always be a need for humans in the economies of the Digital Age. I doubt that the O-ring production function offers job security to humans who find ways to be useful to potential employers. Autor's argument applies to *roles* in production chains and not to their *occupants*, be they human or machine. The traditional way to improve a link filled by a human is to pay to improve the skills of the current human worker or to hire a more-skilled human replacement. This leads to better pay for humans who remain in the production chain. But another way to improve that link in the production chain is to replace human workers with more efficient and cheaper machines. It would be wrong to suppose that human workers will be completely eradicated from a production chain. Our ingenuity helps us to find other ways to contribute. But as we demonstrate the value of these new contributions, we provide a powerful economic incentive for our replacement by the protean digital package. We see the difference between the Industrial Revolution and the Digital Revolution. A foreman in a factory of the early 1800s could feel much more secure in his job than does a human worker who finds a new role in a digital economy production chain. No modified power loom threatened to take the foreman's job.

The principal problem for inductive optimism lies in the capacity of the machines of the Digital Revolution to follow us into new lines of work. Digital technologies are protean in a way that the technologies of the Industrial Revolution were not. A power loom casts the handloom weaver out of a job. But it has no capacity to pursue the weaver's son into his job as factory foreman. Machines capable of mind work can do this. We can always invent new tasks for human workers—there is no reason to think that the supply of potential human jobs is finite. The problem is inventing new tasks that cannot predictably be better performed by the machines of the Digital Revolution. Advances in machine learning will permit machines to do mind work. The protean nature of machine learning means that though we may think of new jobs that seem better suited to the digital package, it will be difficult to prevent digital machines from doing them better than us.

The optimists find little difference between forecasts of doom about work made at the time of Industrial Revolution and today's despairing prognoses about the Digital Revolution. But there's a big difference. The Industrial Age created jobs beyond the imaginative limits of its contemporaries. But the hope that the future would contain many jobs for humans would have found support in the many jobs of those times that could not be done by power looms or any other technology produced by the Industrial Revolution. A sporadically employed handloom weaver would have little difficulty in thinking up jobs that faced no threat from the varieties of automation brought by the Industrial Revolution. His son would be ill-advised to take up handloom weaving. But jobs keeping the books of commercial enterprises, in the army, or in domestic service would predictably continue to be available. No steam-powered machine could do these jobs. The protean nature of computers makes the focus of today's anxiety about technological change broader. The parents of the early decades of the twenty-first century aren't so much worried about children being unable to follow them into the family line of work. They struggle to think of any job predictably on offer in the Digital Age. The existence of many jobs during the Industrial Revolution not under threat from steam power and its allied technologies would have supported confidence about future jobs. The protean nature of the digital package speaks against an expectation of jobs for humans in the Digital Age.

We should understand the limited value of economists' conjectures about the effects of technological progress on current patterns of employment. In chapter 1, we saw that relevant differences between yesterday and tomorrow should reduce confidence in their inductive inferences. The data supporting these conjectures about today's economic trends are uninformative about the changes brought by the novel combination of data and

AI. A very good explanation of trends up until now may fail to take into account changes brought by the Digital Revolution. The economic data about the effects on work of the recent past of automation are uninformative about predictable changes in digital technologies.

Consider the influential account of the effects of technological change on employment from David Autor and David Dorn.[13] Autor and Dorn challenge the received view among economists that technological change favors workers who are more skilled and therefore make better use of new technologies. According to them, this hypothesis of "skill-biased technological change" fails to adequately account for an observed polarization in the economies of technologically advanced nations. There have been increases in employment and wages of the most skilled, but there have also been increases in employment and wages in the comparatively unskilled service occupations. Autor and Dorn define service occupations as "jobs that involve assisting or assisting or caring for others, for example, food service workers, security guards, janitors and gardeners, cleaners, home health aides, child care workers, hairdressers and beauticians, and recreation occupations."[14] A label for an explanation that better explains this polarization is "the task-biased technological change hypothesis."[15] Dorn writes "In the TBTC model, computers do not have a differential impact on workers based on their education levels, but based on the task content of their occupations." Autor and Dorn surmise that automation is responsible for this polarization. Many middle-income jobs are being lost to automation. We've seen how machines can do the mind work of accountants both better and more cheaply. But jobs at the top and bottom are less susceptible to automation.

Dorn highlights the features of hotel work that make it difficult to automate.[16] The cleaning of hotel rooms is repetitive. But according to Dorn, this repetitiveness does not translate into routines that can be programmed into computers. He says, "For hotel cleaning to be a routine job, it would be necessary that the cleaning of one room would encompass exactly the same work steps and physical movements as the cleaning of the next room. But in practice, every guest will leave her room in a slightly different state. Apart from differences in cleanliness, guests can leave towels, pillows, toiletries, pens and many other objects that belong to the hotel in different spots within the room." Dorn continues: "It would be very challenging for a robot to find and recognize all of the hotel's objects, assess their state of cleanliness, and take the appropriate measures of cleaning or replacing them. Compared to humans, robots are often very limited in their physical adaptability, and cannot grip or clean many different types of objects."[17]

If digital machines are to do hotel work, then they cannot do it as humans do. Businesses will need to find different ways to get this hotel work done if they want to automate it.

There are many historical precedents for technological advances disrupting familiar human ways of doing work. Before the Industrial Revolution much production was done in workers' homes. Merchant employers would put out materials to rural producers who would typically work on them at home, bringing finished products back to their commissioners. It would be difficult to see how the Industrial Revolution innovation of the steam engine could have any positive impact on work done at home. Miniaturized steam engines that could be installed into home workshops were beyond the ken of Thomas Newcomen or James Watt. The innovation of the steam engine required that people cease working at home and travel to a factory. We should expect analogous changes to the Digital Age centers of production.

We see significant progress on the dehumanization of work and the workplace to better meet the needs of digital machines in Amazon's fulfillment centers. Writing in *Wired*, Marcus Wohlsen notes that Amazon's fulfilment centers are not organized in ways that make sense to humans. They are not like department stores that subdivide items in ways that meet the expectations of human shoppers. If you want running shorts, then look in sportswear. If you're after a gift for a child's birthday, then seek out the toys department. It would difficult to design a robot capable of searching department stores with the same facility as human shoppers. In an Amazon fulfillment center, items are grouped in ways that take little interest in the way humans group like with like. Their codes are stored in the fulfillment center computers. Wohlsen says "… unlike the Dewey Decimal System, the codes don't signify anything about the category of what's in the cubby. Items are simply shelved where they fit, with identical copies stowed in spots throughout the warehouse to make them accessible to make it less likely a worker will have to travel far to find one."[18] Amazon's computers know exactly where everything is—they have no need for the step in locating tennis balls that goes "These are sporting equipment, so they must be in the sports department." Items' locations make little sense to humans whose first instinct when searching for men's shorts is to seek out the menswear section, but that doesn't matter if Amazon's AIs treat its workers as biological drones. Amazon's replacement of these humans by Kiva robots is an obvious next step. Amazon makes progressively less use of the advantages that its human workers have over Kivabots. Kivabots would be flummoxed by the layout of a department store but do just fine with the inhuman layouts of Amazon's fulfillment centers. The Digital Revolution is turning many workplaces into human-hostile environments much in the way that the warming of the Coral Sea is making it a hostile environment for coral reefs.

The task of programming an industrial robot to respond to the chaos of a recently checked out from hotel room seems truly immense. The challenge for hotels is to change so that their rooms both meet guests' expectations and suit the cleaning robots that enter when the guest leaves. They will look to indulge guests' desires to be randomly and sometimes disgustingly messy, but also explore ways to encourage them to do so in ways that cleaner-bots can cope with. They will use machine learning to find useful patterns in the messes made by many millions of hotel guests. Remember that we need not suppose that machines will do *all* hotel work for there to be major problems for the hotel industry as a source of employment. Imagine that cleaning bots cannot safely navigate the corridors of hotels. Human employees are needed to carry the bots into each hotel room. That leads to fewer human hotel employees and a further blow to the work norm.

Will Humans Always Control the Last Mile of Choice?

We should not place too much confidence in the preservation of human O-rings in the hotel industry. But perhaps we shouldn't be mourning the abolition of low-skilled jobs if the prospects for human high-skilled workers are favorable. According to Autor and Dorn high-skilled jobs are also resistant to automation. Those who might have missed opportunities to educate themselves will get the message that the future will contain very few jobs driving delivery trucks or cleaning hotel rooms. They will acquire management and professional skills and reap the economic rewards brought by high-skilled occupations. According to Dorn a common feature of high-skilled jobs is that they "draw on human ability to react to new developments and problems, and to come up with new ideas and solutions." He suggests that computers complement human workers in these areas, but they are not substitutes for them.

This confidence about the prospects for high-skilled workers finds partial support from Pedro Domingos, the advocate for machine learning whose ideas we have explored in chapters 1 and 2. I say "partial support" because we have seen how Domingos imagines machines outdoing the capacity of the most educated and creative humans to come up with novel responses to disease. The news may be better for highly trained managers than it is for professionals whose education involves internalizing knowledge that is more efficiently and completely acquired by machines learners.

Domingos is confident humans will control the "last mile of choice." He offers the following characterization of the relationship between human deciders and the decisions taken by increasingly powerful machine learners: "The last mile is still yours—choosing from among the options the algorithms present you with—but 99.9 percent of the selection was done by them."[19] The machines will pass their conclusions on to humans who will make the final choice. The role of CEO of choice seems to play toward human strengths. A good CEO lacks the specific skills and expertise of most of her subordinates. These skills and expertise tend to distract from her main role, which is to adopt an organization-wide perspective. The CEO's expertise lies in making choices about the overall direction of the organization. Tesla Motors CEO Elon Musk needs to know a great deal about how electric cars work and could work in the future, but it would be a mistake for him to busy himself with the minutiae of a new design for windshield wipers. He operates principally in the last mile of choice.

Again, a good explanation for trends up until now should not be relied upon as a forecast of the Digital Age. I argue that the last mile of choice is an especially bad place to locate the humans of the Digital Age. It violates a principle of good collective decision-making according to which it is wrong to give less competent deciders authority over more competent deciders. Whether one party can serve as a useful final decider with authority over the decisions made by another party does not depend purely on facts about the competence of that decider. It depends on an assessment of the *relative* competence of the candidate final decider and the deciders over which authority is claimed. This relative competence matters more than any objective assessment of the absolute quality of decisions. It is not enough for someone to demonstrate that they can make very good decisions about the dos and don'ts of heart surgery for them to claim the role of final decider over your procedure. They should demonstrate a capacity for good final decisions at least as good as those over whom they exercise authority and other candidates for the role. This has implications for human deciders in the Digital Age. The need to place humans at the summit of the hierarchies of choice in the Digital Age reflects present bias about computers and an unwarranted belief in human exceptionalism. We should assess their long-term eligibility for that role relative to predicted digital deciders. It's not enough to make better decisions than any other human.

When considered in the long term, the importance of decisions taken in the last mile of choice makes it the worst place for an organization in the Digital Age to locate its remaining human workers. The last mile contains decisions where the advantage of future machines over humans is likely to be especially great. If you had to give human deciders a mile, it should not be the last one. The final mile brings, what is in decision-making terms, the point of no return. Foolish decisions cannot be scrutinized and corrected by a decider that comes later. Suppose Michelangelo had taken on an apprentice sculptor whom he assessed as particularly talentless. He is obliged to offer the apprentice some way to contribute to the forthcoming work. If Michelangelo has any interest in the quality of the sculpture, he should not allocate the last mile of sculpting to the talentless pupil. Mistakes made earlier may not be entirely reparable, but at least *il Divino* can do something to tidy them up. There is no such opportunity if the hack takes charge of the last mile of sculpting, no opportunity for Michelangelo to tidy up a few misdirected blows, cleverly refashioning a botched heroic sword into a discreet dagger.

The qualities that make one human final decider better than another human candidate for that role are unlikely to guarantee success into the Digital Age. One vice of a human final decider is micromanagement—the allocation of too much attention to the minutiae of tasks performed by subordinates. Human managers who seek to micromanage tend to lose sight of the big picture. Digital final deciders will predictably make better final decisions because their micromanagement can inform their all-things-considered resolutions, and vice versa.

We can describe this capacity in a way that is neutral between the mind work of humans and computers. A computer has active memory that includes tasks that it is currently working on. Humans have something analogous—what psychologists describe as a "mental workspace" in which we locate the tasks we are actively addressing. Computers slow down when their active memory gets clogged with tasks. Human decision-makers become less efficient as their mental workspace fills with tasks.

Effective human final deciders conserve their mental workspace by avoiding cognitively bogging themselves down in the detail of a task assigned to a subordinate. Good human subordinates know how to give a final decider the information she needs to make good decisions. They do not presume to pre-empt those decisions, but they are careful to omit details that hamper a final decider's understanding of how the processes they are reporting on relate to other processes in the organization.

The danger of clogged mental workspace is starkly demonstrated in tragedies that have occurred in the cockpit. Planes crash when pilots become absorbed with the problem of how to fix an apparently faulty dial and lose situational awareness, failing to notice a fatal loss of altitude. The ideal final decider in the cockpit must allocate some of his mental workspace to proximal dangers but reserve sufficient workspace to an overall awareness of how the aircraft is flying. This is a difficult trade-off for humans to make. It seems to require split attention in which a pilot attends to a new alert without losing a general sense of how the plane is flying.

An effective human final decider zealously guards her mental workspace. Digital final deciders have another strategy available to them. Their designers can simply give them more active memory. If your computer slows down when you instruct it to simultaneously perform too many tasks, you can boost its memory. For digital final deciders, attention to the minutiae of one problem need not lead to poorer big picture choices. We should not overstate the capacities of today's computers. Computers do not have limitless active memory. Active memory is a precious resource that computer engineers diligently preserve. But we are subject to present bias if we assume that the specific bandwidth limits of today's machines will limit tomorrow's machines.

Some people take great pride in their capacity to multitask—to deal simultaneously with many different steams of information that require distinct responses. Rather than genuinely multitasking, humans serially single-task, rapidly switching their focus from one stream of information to another. This process of switching back and forth leads to significant declines in performance. It explains how a pilot might become so absorbed by an aberrant dial that she fails to notice a precipitous decline in altitude. She neglects to switch her attentional focus back to assessing how the plane is flying. Computers are, in contrast, genuine multitaskers. The number of distinct tasks performed by a computer is limited by its processing power. Giving a powerful computer additional tasks shouldn't lead to any decline in its performance of an existing task so long as sufficient processing power continues to be allocated to it.

Consider the pragmatic trade-offs in decisions about what information to relay to a human final decider in the context of the jetliner cockpit. The sensors of a modern jetliner have access to a great deal of information about what it is doing, and what is going on outside of it. Part of good cockpit design is to work out how to display this information to the pilot in a way that respects human cognitive limits. There's only so much information available to aircraft's sensors that can usefully be presented to a human flight crew. The design of aircraft instruments and dials must respect the biological limits of human pilots. There are probabilistic thresholds built into the design of cockpit instrumentation. Good cockpit design omits information assessed as having a low probability of making a difference to the flight. Once we disabuse ourselves of the twin distortions of present bias about digital machines and the belief in human exceptionalism we can appreciate possibilities for final deciders with much greater bandwidth. We should open ourselves to the realization that information that fails to exceed the probabilistic threshold to be worth reflecting in the dials and instruments of a cockpit occupied by a human flight crew could be worth passing on to a final decider with much greater bandwidth. This final decider could make use of data about the angle and velocity of winds that are rightly deemed not worth passing on to human flight crew. A tiny difference in the velocity or angle of a wind is unlikely to make a difference to flight safety. But, every now and then, it will. If there were no limits to the calculations and calibrations that can be introduced into a pilot's active memory, then this difference should be considered. We are unlikely to ever have a computer with infinite active memory. But Moore's Law and related generalizations suggest futures in which computers get ever closer to this ideal limit. So long as we allocate the last mile of choice to human pilots there seems to be a limit to how much safer we can make our planes. It's safer to board a plane whose pilot's active memory is sufficiently capacious to consider even improbable threats than it is to board a plane whose pilot's active memory is more limited and has been forced to disregard threats that do not exceed a certain probabilistic threshold. There are many such individually insignificant threats. When these threats are considered in combination, we see the risks human final deciders in the cockpit subject us to.

What goes for the last mile of choice in the cockpit goes also for the last mile of choice elsewhere. Consider the predicament for the human CEOs of the Digital Age. Attributes such as guts, intuition, and insight into the motives of others tend to feature prominently in today's accounts of business success. Today's business leaders successfully adapt cognitive hardware designed for the Pleistocene to the combination of opportunities and threats presented by early twenty-first century business environments. Perhaps we should be impressed by the cost-cutting genius of General Electric's Jack Welch and the iconoclasm of Apple's Steve Jobs. But we should consider these successes in a context that includes the epic fails in which human business leaders made ruinously bad choices. Jobs has his counterpart in John Sculley, the Apple CEO from 1983 to 1993. Sculley's poor understanding of Apple's products prompted him to sack Jobs. Think of the vast quantity of data that a business AI will be able to draw on to make its recommendations. It will draw on the totality of available information about the stock prices, business cycles, historical records of good and bad corporate takeovers, and so on. Machine learners will search for patterns in this data and seek to apply what they discover to future decisions. It seems plausible that it will improve on humans who draw on the small subset of the available data that their "guts" direct them to.

The all-things-considered character of final decisions is likely to be a strength of future AIs when compared with selective attention and imperfect multitasking of human final deciders.

It's important to not understate the implications of the predicted domination of machines over the last mile of choice. Some writers hope that there will still be a role for humans over this decisive period of choice. For example, McAfee and Brynjolfsson allow that human diagnosticians could have some part to play in the detection of disease in the Digital Age. Speaking of a medical diagnosis offered by a future AI, they say, "It might still be a good idea to have a human expert review this diagnosis, but the computer should take the lead."[20] So long as humans play some part in the final mile then we have some latitude to describe our contributions as the most important. CanceRx is incapable of objecting that some human "expert" who spent much of the time watching it do its thing gets to claim the credit and the Nobel Prize for a radical new approach to leukemia. CanceRx can make a mechanical genuflection toward the need of a human decider to be "the man" or "the woman." But the point is not merely that, when considered objectively, human contributions will be less important. If the gap between human capacities as final deciders and the capacities of the machines is so great, then humans should play no role whatsoever. It doesn't matter how many Wikipedia pages on cardiac function you've browsed, when you're watching open heart surgical procedure it's probably best that you keep your thoughts about which incision to make next to yourself. Your input is likely to make the procedure go worse, not better. The same warnings apply to the human expert reviews of the diagnoses of Digital Age medical AIs. When autopilots get manifestly better at flying planes than human pilots we should not grant humans any opportunity to override the autopilot's choice. The cockpit doors of Digital Age jetliners should be both terrorist proof and human pilot proof.

There is a defense of human final deciders that acknowledges our inferiority in this role. We could accept heightened risk as part of the price of dealing with organizations controlled by humans. We knowingly engage in many dangerous activities. Reflective passengers understand that they do not board a commercial jet with an absolute guarantee of safe arrival. They understand this risk and judge it acceptable. They might accept that giving employment to a human pilot—the human final decider cockpit—justifies a little bit of additional risk of violent death. If we happily judge the pleasure of seeing Siena sufficient to justify boarding a commercial jet, then why shouldn't we accept that it's worth a bit of extra risk to keep a human pilot in gainful employment.

This book draws a distinction between the digital and social economies. The digital economy centers on the value of efficiency. We care about efficiency in the social economy too, but we permit our preference for humans to sometimes lead us to prefer less efficient arrangements involving humans to more efficient arrangements without humans. We are prepared to tolerate inefficiency when we get to enjoy the benefits of interacting with humans. But there are activities in which there don't seem to be compensating benefits from interacting with humans. We treat occasional errors by our pilots differently from the errors by our baristas. If we got to interact with our pilots we might judge the additional risk of death as warranted.

But the threat of terrorism has increasingly isolated them from us. Where there are no compensating benefits from humans we will be increasingly intolerant of the additional risk of dealing with humans.

We should take the long view of our assessments of risk. There is no objective feature of human physiology or psychology that makes a certain level of risk acceptable and a slightly higher level of risk unacceptable. We judge risk relative to other things that we do. Simply put, life has gotten safer over the past centuries. This has led us to find activities that were formerly safe enough to be too risky. Many people today find the risks associated with smoking to be unacceptable. We should expect, that were they offered cigarettes, people in the Middle Ages would make different assessments. They would be able to look at the health statistics on smoking that deter many of us, consider these in the light of features of their daily existence—awaiting the next visitation of the Black Plague, getting forcibly inducted into a neighboring lord's army, and starving when a crop fails. They would be likely to judge an increased risk of cancer from smoking to be entirely acceptable.

The relativity of assessments of risk has implications for the human final decider in the cockpit. Suppose we continue to make improvements to aspects of our lives outside of flying. Advances in digital technology produce improved treatments for many of our diseases. Today's travelers happily tolerate the risks associated with human pilots. But tomorrow's passengers predictably won't. They will view consenting to travel in a human-piloted jetliner as hideously reckless.

A Conjecture about the Labor Market of the Digital Age

The threat to the work norm from the digital package is sharpened when we give it some economic context. Suppose that humans follow Autor's advice and seek to turn ourselves into digital economy O-rings. We will attempt to justify high rates of pay by pointing to the great importance of our contributions. These efforts could work in the short term. However, the long view exposes this strategy as self-defeating. What's essential to the economy is the O-ring role—if these O-rings are truly essential then Digital Age economies without them will crash or explode. But this reasoning says nothing about who or what fills that role. The human worker in a Digital Age production chain should understand that the better she does, the stronger the economic incentive to design the digital machine that will fully or partially replace her. The human mind workers remaining in the production chains of the Digital Age should feel like hunted foxes, fleeing from temporary refuge to temporary refuge just ahead of the pursuing hounds. The economic argument against human work suggests that refuges will be temporary. The only way to feel truly safe is to convince those finding new applications for the digital package that the economic value of what you do for your living is negligible.

Earlier we saw Autor and Dorn's explanation for the polarization of the Digital Revolution workforce. They explain this by pointing to the differential impacts of automation. Jobs at the top and bottom are in general harder to automate than are jobs in the middle. I suggested that good explanations of the present will fail to predict the future if conditions change. We should expect reorganization of workplaces to make jobs at the bottom easier to automate. The creativity and insight celebrated by business biographies must be judged against the pattern-finding prowess of future machine learners. The same digital superintelligence that we could unleash on cancer will predictably be directed at the challenges of business.

I conjecture that the polarization that Autor and Dorn present as a feature of today's labor market will characterize the labor market of the future, but for purely economic reasons. The worst remunerated jobs will last longest simply because those who fill them will work for less. Hotel workers are poorly paid. If they were paid at the same rate as accountants, we might expect to see an increase in the pressure on their jobs from automation. Automation is characterized by steep one-off costs associated with introducing the systems that will do the work of human workers, followed by much lower costs of maintaining and occasionally updating the machines. It differs from the high ongoing costs of paying for human employees who expect periodic pay rises. The strategy of working for less reduces these ongoing costs. Poorly paid human workers should last longer in the workforce before improvements in automation make the economic case for replacing them irresistible. Accountants go before hotel workers, but eventually even the poorly paid can no longer compete on price.

I suggested that the digital package can more easily take on the work challenges of those at the top than supposed by Autor and Dorn. Machine learners should improve on the decision making of captains of industry.

In late 2015 Mark Zuckerberg stated "My personal challenge for 2016 is to build a simple AI to run my home and help me with my work." He expected that this AI would be a bit "like Jarvis in Iron Man."[21] The suggestion that Zuckerberg's version of Jarvis will "help" him seems to arrogate to himself more than his due for the business triumphs of a future Facebook helmed by a combination of Zuckerberg and Facebook-Jarvis.

Zuckerberg is likely to make poorer business choices than will a Facebook-Jarvis trained up on the totality of information about stock prices, business cycles, and historical records of good and bad corporate takeovers. His choices should be viewed as we will predictably view the recommendations of human experts in the genetics of cancer when the alternative is CanceRx. But there is a reason that we should not expect Zuckerberg to yield his position at Facebook's helm to a machine. He enjoys running the company he founded. There's a good chance that considering which businesses are potential acquisition targets for Facebook puts him into a flow state. To put this another way, Zuckerberg does not want to become a rentier—an individual who lives off rents yielded by his assets. We know this because he could become a rentier right now. Zuckerberg has amassed sufficient wealth to spend his life flitting from luxury resort to luxury resort. He values his own agency. Even when it is clear that he has nothing of value to add to Facebook-Jarvis's recommendations he will enjoy exercising control over the last mile of choice over which businesses make suitable Facebook acquisition targets. Zuckerberg and his inheritors have the money to exercise agency in the economics of the Digital Age even when a sober evaluation of their competence compared with the competence of Facebook-Jarvis tells them they should become rentiers and split their time between luxury ski resorts and tropical island paradises.

When a poor person's job is better done by a machine he finds himself facing penury. Things are different for the proprietor of the machines that render the poor person redundant. Zuckerberg values his own agency. He can pay for the luxury of exercising it when digital deciders are clearly superior. When he does this, he acts like a spoiled medieval princeling who chooses to lead an army into battle when the grizzled low-born veteran would do a better job. He gets to lead the army because he's the prince.

In chapter 5 I explore the prospects for spreading this celebration of agency more broadly than those who have the money to indulge the illusion that they can out-think the machines. But I conclude this chapter with a philosopher's take on how best to approach disputes about what the future will bring.

Gaining Philosophical Perspective on the Dispute between Optimists and Pessimists

On one side of the dispute about the role for humans in Digital Age economies, we have the optimism of the economists that places faith in the ingenuity of humans to find work that is both productive and therapeutic. They support this optimism with an impressive inductive argument that points to many other cases in which jobs seemingly materialized out of nowhere. They can cite a history of cases in which despair about the future of work was followed, after a painful delay of technological unemployment, by jobs that we didn't, and indeed couldn't, have imagined before their arrival. We compare the jobs we lost with those we gained and consider ourselves ahead on the deal. We should expect to find new ways to make ourselves useful as our machines get more and more powerful. On the other side, we have the pessimists who point to the protean powers of the digital package. When combined with the economic argument against human work we should expect the package to promptly eliminate newly discovered sources of productive and therapeutic work. The idea that we might place ourselves at the top of decision-making hierarchies wrongly arrogates to ourselves a capacity to make better decisions than AIs. The protean powers of the digital package permit it to fill any new economic roles.

Should we be optimists or pessimists about the economic value of human agency in the Digital Age? There is some evidence that a bias toward optimism is beneficial for individuals. People who are clinically depressed tend to have more realistic assessments of their social standing. If optimism bias is the price for navigating the social world with confidence, then it seems a price worth paying. When it comes to the Digital Revolution, however, there are costs in the resolution to "look on the bright side." We are better advised to approach the Digital Age's uncertainties as pessimists. It may be good for individuals to have an optimism bias, but a pessimism bias often works better for us as collectives. Collective pessimism is an essential hedge against the Digital Revolution's turning out worse that we might expect.

What is the rational way to respond to a disagreement about the future? The traditional way is to try to work out who is right. You should earnestly commit yourself to the task of working out which side of this debate has the strongest arguments. This approach works very well for academic seminars. But it doesn't work as well when participants in those debates emerge from the seminar room and start offering advice to those potentially affected by the topics of their debates. There are dangers in premature resolution of the debate between optimists and pessimists about the work norm in the Digital Age. Neither the optimists nor the pessimists possess crystal balls or De Lorean sports cars modified to permit time travel. Under these circumstances, we should place greater emphasis on an attitude toward the future that best insures us against possible misfortunes.

You should take the same approach that you take to purchasing fire insurance for your dwelling. My house in Wellington has stood for over one hundred years without burning down. I'm quite confident that it won't burn down in the near future. It has an open fireplace that we scrupulously avoid using even on the coldest winter nights. We nevertheless have an insurance policy against fire and are up to date with our premiums. If my house never suffers fire damage, then this particular expense seems a waste of money. The money would have been better spent on meals out and movie tickets. But this is the wrong way to think about insurance. It can be rational to insure against unlikely events if you think that the occurrence of these events would be a disaster. When I consider purchasing a fire insurance policy I consider how likely a fire is to occur. If I assess the probability of that event at zero, then I shouldn't purchase. But suppose that I decide that there is some non-negligible probability of a fire. I consider the cost of the policy and how bad a house fire is likely to be. If the premiums are expensive relative to the value of my house I don't purchase a policy. But if those premiums are sufficiently cheap relative to that value I do.

It's useful to take this insurance mindset into the debate between the optimists and the pessimists.[22] We can view the claims of the pessimists as constituting a variety of insurance against a future in which advances in digital technologies reduce the economic value of human agency to such an extent that humans cannot find work. If the optimists are right, then this imaginative effort may seem wasted. Fantastic jobs that we failed to imagine materialize. Instead of working out how to respond to the threat of

Digital Age joblessness, creative people might have invested their imaginative labor elsewhere. We can count the cost of this effort in terms of the symphonies and computer games that could have been created were people not so worried about how to respond to a threat that never materialized.

Consider Aesop's fable about the boy who cried wolf. The shepherd boy becomes bored watching his sheep, so he decides to amuse himself by crying "Wolf! Wolf!" The villagers dash out of their dwellings to drive the wolf away and find there is no wolf. The boy repeats this a few times. Finally, there is actually a wolf. The boy's calls of alarm elicit no response. The wolf scatters the herd, leaving the boy in tears.

You can understand the parable's message emphasizing the great importance of telling the truth. If the shepherd boy didn't have such a track record of fibbing, people would have listened to him when he warned of a real wolf. But there's also a message from the parable for those who find themselves cast in the role of the villagers. When dealing with claims about which we cannot be certain don't place too much confidence on past nonappearances by wolves. The fact that the wolf has not appeared after each earlier warning seems to strengthen the inductive case for the wolf never turning up. But the fact that we feel more confident about the nonarrival of the wolf with each past nonarrival does not obviate the importance of maintaining a look out if it's possible to do so in ways that do not harm other vital undertakings. Boys should not make up stories about wolves, but equally villagers whose flocks are threatened by wolves should not suppose that the nonappearance of wolves up to this point means that we shouldn't worry at all about future wolf attacks. Shepherd boys who persist in making stuff up should be scolded. But equally if it's relatively easy to scare the wolf away, the villagers should rush to defend their precious herds whenever they hear the cry of "wolf." We could supplement the traditional message of Aesop's fable—"If you persist in making things up, people won't believe you in the future" with "If someone warns you that something bad will happen, think about the cost for you of doing something about it. If the cost is low, don't be overly influenced by inductive arguments against doing anything."

Some optimists about technological progress point to the "Great Horse Manure Crisis of 1894" to illustrate the foolishness of pessimism about the future.[23] Each day in the period leading up to 1900 saw more than 50,000 horses ferrying people and stuff around the streets of London.[24] These horses generated huge volumes of manure. A quote widely attributed to the *Times* of London in 1894 considered how much manure horses generate and made the forecast: "In 50 years, every street in London will be buried under nine feet of manure."[25] We today look back and wonder how the alert observer could have missed the imminent arrival of the automobile, which would radically reduce the number of horses in London and turn the Great Horse Manure Crisis into a problem that required no solution. No one likes to be laughed at. The fact that it's posterity that's laughing at you does little to lessen its sting.

The insurance mindset I recommend here suggests a different evaluation of the Great Horse Manure Crisis. The Londoners of the 1890s lacked time-travel equipped Hansom cabs. Perhaps some people particularly well informed about the automotive advances of Karl Benz in the 1880s and 1890s might have ventured conjectures about the potential for his inventions to transform our cities. But they couldn't have been certain. What if the fumes produced by Benz's automobiles proved toxic, preventing their introduction to London? The insurance mindset asked how much it would have cost us to do some thinking about what to do about London's increasing quantities of horse manure. Thinking about ways in which London's streets could be periodically cleansed seems like cheap insurance against an uncertain future.

We evaluate insurance policies in not as accurate or failed predictions about the future, but instead in terms of their cost. Insurance cover against improbable misfortunes can be worth purchasing if it is cheap. Some imaginative effort seems a cheap premium for insurance against a possible future in which the work norm does not survive into the Digital Age. If the optimists turned out to be right, then we had nothing to worry about. But that doesn't mean that it was wrong to acquire insurance against the bad outcomes described by the pessimists simply because they didn't materialize. If we do not put significant thought into how to confront a future without jobs, then we are underinsured for the Digital Age. We can hope that the optimists are right much in the way that someone who forgoes fire insurance hopes that her house will not catch fire.

Suppose the prediction that the Digital Revolution poses a threat to the economic value of human agency not posed by earlier technological revolutions turns out to be false. After a "temporary phase of maladjustment" ends, the new technologies create new challenging and rewarding jobs that could not have been imagined before their arrival. Each new advance demands new varieties of human O-rings. These new jobs better promote flow than the jobs that went out of existence. People who perform them look pityingly at those stuck with the drudgery of categories of work that thankfully no longer exist. Should we regret the doom saying about the death of work and the alternative scenarios that it forced us to consider? Not if we view that thinking as insurance against a future in which each new economic role is better performed by machines. How should I think about the many years of premiums I have paid insuring my house against fire? There's a sense in which this is wasted money. But it's not if I view the policy as offering me protection into an indefinite future. This is the way we can think about our preparation for a future in which the Digital Revolution destroys many jobs and creates some new jobs, but too few to sustain the work norm. We can be thankful that the Digital Revolution left the work norm intact, but also be grateful for the preparation for any future technological revolution that does end human work. The work norm might flourish into the Digital Age. But you should be much less confident that this success will be repeated for future technological revolutions. Will the work norm survive into a technological age centered on the immense productive powers of quantum mechanics?

Keeping up with your insurance premiums offers you protection into the indefinite future. So it is with some creative thinking about new possibilities for human work. In the chapters that follow I direct this thinking away from the suggestion that we might be better than machines at the tasks they are designed to do. In chapter 2 I argued that the quest for machines with minds takes a distant second place to the quest for machines that do mind work. Secure work for humans depends on the facts about us of which we are justifiably most proud. We have minds. The most powerful mind workers of the Digital Revolution will not be members of the mind club.

Concluding Comments

This chapter explores the debate between optimists and pessimists about the economic value of human agency. Optimists expect the work norm to survive into the Digital Age. Pessimists allow that the Digital Age will contain jobs done by humans but doubt that these will suffice to maintain work as a norm for humans. The optimists bet on human ingenuity. We've always responded to past challenges by finding new ways to be useful. Pessimists counter by pointing to the protean powers of the digital package. I suggest that we should place greater emphasis on what the pessimists say. We should view pessimism as valuable and cheap insurance against a jobless future.

Preparing for a world in which the pessimists are right is relatively cheap insurance. It requires people to think seriously about how to respond to a future in which the Digital Revolution destroys many more jobs than it creates. This pattern places the work norm under threat. If the optimists are right, there will be some wasted imaginative effort. We could have just relaxed and enjoyed fantasizing about the new jobs done by our children and grandchildren. But that cost is trivial compared with what we lose should we march confidently into a Digital Age in which many jobs are destroyed but none of the jobs that we were hoping for ever actually arrives.

In the chapter that follows I consider a different way to value human agency. We should look to the Digital Revolution to create new categories of jobs for humans to fill. We should accept that any new economic roles will be better filled by machines than by humans. We should instead consider precisely what we value about interacting with each other.

5 Caring about the Feelings of Lovers and Baristas

Preparing for a future in which the digital package eliminates many jobs and fails to generate many to replace them is good insurance against the uncertainties of the Digital Age. We must look beyond the digital package for jobs to replace those that it destroys. It is a mistake to assume that the digital package will, after a brief interlude of technological unemployment, create better jobs to replace those it destroys. We should expect the digital package to create new economic roles. Humans may be able to fill these roles. But the protean powers of the digital package should lead us to expect that these roles will be better filled by machines than by humans. The money that would fill the pay-packets of these human workers creates an incentive to make machines that are both more efficient and cheaper.

This chapter makes a case for a category of social jobs grounded in our interest in interacting with beings with minds like ours. Humans prefer to interact with members of our particular chapter of the mind club. This vision of how we could inhabit the Digital Age requires that we reconsider what we value about work. I argue that we should create societies organized about social-digital economies. These economies feature two different streams of economic activity, centered on two different kinds of value.

The principal value of the digital economy is efficiency. We will assess potential contributors here in terms of the efficiency with which they produce outcomes that we value. If we characterize the purpose of work in terms of its outcomes, then we cannot long resist the machines.

The principal value of the social economy is humanness. The rewards here flow to beings with human minds, beings capable of feelings like ours. We take pride in our distinctively human chapter of the mind club and enjoy interacting with other members of it. We value efficiency here too.

But we sometimes prefer less efficient arrangements with humans to more efficient arrangements without humans. Turing fantasized about machines with minds like ours. The pragmatic interest in AI that has given us machine learning directs us to enquire after the opportunity cost of making machines capable of small talk. The machines that really power the Digital Age will lack the all-round combination of mental abilities that we equate with being human. We will rightly reject the inefficiencies of humans when they stray into parts of the economy that emphasize the skills of the computer. But we should have the courage to reject digital technologies when they trespass on distinctively human domains. When we do so, we cite our strong preference for outcomes produced by beings with human minds.

The social-digital economy is not a forecast. We could easily realize some version of a digital dystopia in its stead. In one version of this digital dystopia, efficiency rules. Humans are forced to compete on price with more efficient machines. Eventually they succumb to the economic argument against human work. I offer the social-digital economy as an especially attractive way for our species to inhabit the Digital Age. It will require some effort and attitude adjustment to achieve it. We hear much about the wonderful toys and apps of the Digital Revolution. We need a matching social revolution—social as opposed to a social*ist*—that gives human workers displaced by the Digital Revolution opportunities to work in the social economy. The Digital Revolution promises the technologies of science fiction. The social revolution could restore some of the gregariousness that was a feature of our pre-civilized past.

What Is It Like to Love a Robot?

The defining features of jobs on the social side of the social/digital divide are human relationships. Many of the most important relationships of our lives occur outside of the context of work. These include relationships with romantic partners, children, and friends. But many important relationships are made and play out within the context of work. We enjoy distinctively human relationships with our employers, work colleagues, and those we employ. We form such relationships with those to whom we provide services, and those who provide services to us. We may place greater importance on relationships that we form outside of work. But that does not prevent relationships that occur in the context of work from mattering too.

The poor performance of machines in jobs centered on relationships is most apparent when we think about romantic relationships. A theme of recent stories about artificial intelligence is that robots make disappointing lovers. In an episode of the TV series *Black Mirror,* Martha replaces her deceased boyfriend Ash with a synthetic look-alike programmed with a personality assembled from Ash's extensive online contributions. She is initially impressed by the duplicate's performance in the sack—human Ash is presented as a somewhat disengaged lover. When Martha asks synthetic Ash how he manages it, his answer—"Set routine, based on pornographic videos"—seems unsatisfactory. It's all downhill from there as Martha tries and fails to elicit from synthetic Ash the kinds of responses that would come from a human lover. The episode ends with synthetic Ash being stored in the attic. There he uncomplainingly awaits sporadic social interactions with Martha and her daughter on weekends and birthdays.[1]

There are aspects of lovemaking that favor the value of efficiency. Synthetic Ash outperforms human Ash in these respects. He lasts longer than 30 seconds and facilitates more orgasms. His mode of programming suggests he could have a *Fifty Shades of Grey* setting that Martha might save for special occasions. These aspects of lovemaking clearly matter. But humanness matters too. Martha wants a lover who does more than just a very good job of going through the requisite motions. She wants to be with someone who can reciprocate her feelings. This, synthetic Ash can't do.

It's possible to imagine improvements to synthetic Ash. His principal deficiencies are in areas in which there is especially rapid progress. His programmers will predictably come up with more human-like continuations of conversations that open with "I love you." Perhaps his romantic programming will achieve a state of virtuosity that makes him hard to distinguish from human lovers. That is something we might expect given the rapidity of progress in artificial intelligence. But this misunderstands Martha's complaint about synthetic Ash. She is not really complaining about his responses. Rather her complaint is about what she strongly suspects is an absence behind those responses. She doubts that there are any genuine feelings and emotions behind his romantic affirmations. Viewed from this perspective, none of the improvements of Ash's programming suffice to transform synthetic Ash from a being with a mental life no different from that of a laptop computer into a being with a mental life like Martha's. Suppose that Ash was a human being who had survived an accident that had damaged the parts of his brain that directed his romantic behavior. Suppose he was attempting to relearn how to make love by obsessively watching pornographic movies. This pattern of behavior is far from the romantic ideal. But Martha could be confident that this version of Ash has a mental life that resembles hers in significant ways. There are real feelings behind his declarations. Real feelings guide his actions.

Consider the Hollywood reflection on robotic romantic partners in the 1975 film *The Stepford Wives*. In this movie, the women of an idyllic Connecticut town have been replaced by robots that care about nothing beyond housework and attending to their husbands' needs. It's hard to tell what is more chilling about this scenario—the idea that the wives might lack distinctively human mental lives, or the idea that the men might take so little interest in the inner mental lives of their wives that they would consider this arrangement an improvement.

I cite the countless love song lyrics that refer to feelings as evidence for this interest in the feelings of our romantic partners. These range from Whitney Houston's request "I Wanna Dance With Somebody (Who Loves Me)," the Bee Gees' query "How Deep Is Your Love," and the fears of The Righteous Brothers that "You've Lost That Lovin' Feelin'."

The philosophical term of art for this feature of Martha that seems absent from synthetic Ash and the Stepford wives is *phenomenal consciousness*. The more familiar term is *feelings*. Put another way, there's something that it's like to be Martha. We worry that there's nothing that it's like to be Ash or the Stepford wives. We are members of the mind club.[2] Our particular chapter of that club is defined by distinctive human thoughts and feelings.

Don't expect a detailed philosophical theory about the nature of feelings here. It is, however, useful to gain some perspective on the seemingly intractable philosophical problem of phenomenal consciousness. When considered in the long view it seems to come from a long-standing desire to find reasons that might justify our specialness. Some religious believers claim that what makes humans different from the rest of nature is our possession of immortal souls. They would deny these to even the most powerful machine learner. According to some, these souls were infused into us by an almighty god. This story about human specialness is less influential in sections of society in which there has been a decline in religious belief. The theory of phenomenal consciousness presents as a replacement for immortal souls amenable to those who lack religious belief. There's something that it is like to be a human being, but nothing that it is like to be a tree or a machine learner. The evidence for the existence of immaterial phenomenal consciousness comes not from holy books but from the data of our experience. In the most historically influential of these arguments, René Descartes noted that our access to and experience of our own minds seems to differ in fundamental ways from our access to and experience of physical objects.

This chapter's engagement with phenomenal consciousness differs from the philosopher's traditional way of engaging with it. The traditional philosopher's question about conscious robots is: "Is there something that it is like to be robot?" Opponents of the possibility of conscious computers identify aspects of the ways in which computers process information that may make this processing crucially different from human conscious thought. They argue that the impressive computational achievements of future computers require no phenomenal states. Defenders of the possibility of conscious computers argue that there are no good grounds to deny them conscious states. As we have seen, computers have demonstrated an impressive ability to achieve cognitive tasks in which humans formerly had a proprietary interest. They beat us at chess and at *Jeopardy!* Why shouldn't improvement in programming and computing power soon produce machines with feelings?

Our starting question switches focus. It takes a second-person perspective on conscious computers. We ask: "What is it like to love a robot?" Doubts about the robot's capacity for conscious experiences make a difference to our experience of being in a romantic relationship with it. We approach uncertainties about the truth of philosophical propositions differently when they make a crucial difference to our lives. When it comes to the relationships that matter most to us, the suggestion that there might be nothing that it feels like to be your romantic partner is terrifying. A romantic partner is supposed to be much more than a sex toy that can also make coffee and pick the kids up from school. This second-person perspective changes how we feel about computers and phenomenal consciousness. Our interest is not first and foremost in the truth of a philosophical proposition. Rather it's in a feature of our lives that most of us place great importance on.

Suppose that you follow the philosophical debate about conscious machines. You find and study the best presentations of the arguments for and against the consciousness of computers. Early on in your investigation you encounter the argument of the philosopher John Searle that computers are incapable of thought, and by implication of conscious thought.[3] As we saw in chapter 2, Searle's famous Chinese Room thought experiment supports the conclusion that even the most sophisticated program neither requires nor generates genuine thought. A quick Google search will reveal many philosophical responses to Searle. Some philosophers think that Searle exaggerates the differences between brains and computers. It's difficult to see how computation might generate conscious thought, but it's also difficult to see how the firing of neurons and the adjustment of synaptic action potentials could do this. In the latter case consciousness seems to be a kind of emergent property that comes into existence with sufficient neuronal and synaptic activity of an appropriate kind. The very complex computers of the future could also generate that emergent property. This is a breathtakingly fast summary of a very involved philosophical debate. Here I am less interested in the details of the debate itself than in making a high-level observation about the practical implications of philosophical conjectures. What should the fact that a very smart philosopher thinks that your cyber-lover who seems attentive to you, is incapable of the barest, most minimal thought, mean for you?

There is a difference between the philosopher's dispassionate engagement with the questions about phenomenal consciousness and the way we tend to engage with questions about the feelings of our lovers. Suppose that you decide that the arguments in favor of the possibility of conscious computers are, on balance, more persuasive than the arguments against. You pronounce yourself a believer in the proposition that computers can be conscious. But philosophical debates are typically not eligible for decisive resolution. This is reflected in the split among informed, intelligent philosophers on the questions of whether a human brain is a prerequisite for phenomenal consciousness. It's helpful to think here in terms of credences, or degrees of belief. A credence of 1 in a proposition indicates absolute confidence in its truth. A credence of 0 in that proposition indicates absolute confidence in its falsehood. Disagreement among those best informed about whether digital machines can have feelings make it rational to avoid either of these extremes. You might conclude that Searle's arguments are, on balance, slightly more persuasive than the many responses. A credence of 0.7 is a rational way to reflect that assessment. It reflects your recognition that there is a real chance that Searle is wrong. Or you might conclude that the arguments of Searle's opponents are, on balance, more persuasive. A credence of 0 in Searle's conclusion misrepresents this considered assessment. A credence of 0.3 should suffice for you to present yourself in a philosophy seminar as someone who rejects Searle's conclusion. You stand ready to adjust that credence up or down depending on any subsequent arguments that you encounter.

Now consider your second-person interest in the phenomenal consciousness of machines. Suppose you discover that the head of your life partner contains not a human brain, but a powerful digital computer. You would presumably be horrified to learn that your significant other not only did not reciprocate your feelings but was not the kind of being who ever could. The idea of a lover who goes through all the romantic motions, performing actions that are perfectly suggestive of feelings but for whom its entirely dark inside, is the stuff of nightmares. The suggestion that the arguments in favor of computer consciousness are on balance slightly more persuasive than those against does not fully meet this worry. Suppose that being generally persuaded by the arguments of advocates of machine consciousness translates into a credence of 0.7 for the proposition that digital machines can be conscious. This might suffice for you to join the philosophical debate on the side of the defenders of conscious computers. But its second-person implications are chilling. It translates into a 0.3 credence in the proposition that there is nothing that it is like to be your lover. As anyone who has ever placed a bet should know, 30 percent of the bets placed on outcomes that have a 70 percent probability of occurring end up losing. When it comes to those you love, you want to hear something better than "The philosophical arguments in favor of the phenomenal consciousness of robot romantic partners are, on balance, slightly more persuasive than the arguments against." You should always be worried when a scan of your significant other's head reveals not a brain, but instead some densely packed circuit boards.

We can render this reasoning in more emotionally vivid, less philosophically abstract terms. Suppose someone were to tell you that there was a genuine possibility that the fully human love of your life, despite protestations to the contrary, actually felt nothing for you. He or she is just going through the motions, exclusively focused on the material benefits brought by the relationship. The love of your life would celebrate your premature passing, performing the grief act just long enough to materially benefit from your will. Further suppose that this possibility is accurately rendered as 0.3. This means that it's more likely than not that the love of your life has feelings for you that resemble yours for him or her. There's a 70 percent chance that he or she is sincere and a 30 percent chance that, deep down, his or her feelings are a combination of contempt and indifference. I suspect that this would be very bad news about your relationship.

We should distinguish this reasonable doubt about the experiences of robots from a species of extreme skepticism whose appeal precipitously declines outside of the confines of academic philosophy. The "problem of other minds" is a philosophical perennial. My capacity for introspection makes me very confident that I have a conscious mind. But I have no such introspective access to your conscious mind. How can I be sure that you have one? Might there be nothing that it's like to be you? Could the world be populated by one conscious being—me—and a few billion mindless zombies in human form?[4] I have nothing further to say about this philosophical skepticism beyond saying that our confidence in the consciousness of other members of our species should be greater than our confidence in the consciousness of even very sophisticated robots. The same neurological hardware that seems to generate consciousness in us also exists in the heads of our human romantic partners. It's the basis of our confidence that dogs that act like they are in pain are in fact experiencing pain. The contents of their heads are similar enough to ours for us to be confident that there is more happening there than a computer's error signal when you persist in pressing the wrong key. If scans of your lover's head reveal innards that are little different from those of a MacBook laptop then you should be worried. Doubts about the reality of his or her mental life should increase. Someone who claims to seriously doubt that the patently human biological brain in your head can generate consciousness should be dismissed as a philosophical crank.

There should be a bias for human lovers well into a future in which sex robots manage to muster human-like answers to "I love you." It's rational to carry a bias for human lovers into this future. Suppose that you have a choice between a human lover and an artificial lover that is the product of a mature artificial intelligence. It reliably produces human-like responses to inquiries. The probability that a programming glitch will lead him or her to respond to some romantic entreaty of yours with "Does not compute!"

is about the same as the probability that your biologically human lover will be unable to respond appropriately because he or she is having a stroke. If you select the first option, then your choice is likely to have been influenced by the belief that humans are special. You may strongly suspect that the machine version has feelings. You might accept that some very intelligent philosophers find no grounds to distinguish between the two potential lovers. But you'll acknowledge that this is a quite serious thing to be wrong about. In affairs of the heart it's best to go with the neurophysiology that you know from your own case can produce conscious feelings. The possibility that the individual that you spend much of your life loving not only doesn't feel the same way about you but is incapable of doing so is, for most of us, profoundly chilling.

I have presented a bias toward romantic relationships with humans and away from relations with artificial substitutes as a rational response to uncertainties about robot experience. There seems to be an emotional grounding for this bias too. Japanese roboticist Masahiro Mori has described an uncanny valley that seems to characterize our responses to artificial beings.[5] We tend to experience a sense of unease when interacting with a computer-generated character or robot that very closely resembles but is nevertheless distinguishable from a human being. The uncanny valley has been an obstacle to Hollywood's attempts to produce animated versions of humans. The 2004 movie *The Polar Express* contained animated characters that looked very like human characters but were nevertheless distinguishable from them. The result is described by the CNN.com movie critic Paul Clinton as "creepy." Clinton says the movie's human characters look "soul dead." The result is that "*The Polar Express* is at best disconcerting, and at worst, a wee bit horrifying."[6] Among the obstacles for CGI characters as objects of empathy or sympathy are their eyes. When we converse with other humans a significant amount of our attention is directed toward their eyes. The white sclera of human eyes seems to have evolved to perform a social signaling function. We have a strong evolved interest in what other humans are attending to. When the eyes of a computer-generated character behave in ways that don't seem human we immediately suspect a counterfeit. The loose resemblance to us of the *Star Wars* protocol droid C-3PO prompts no sensations of angst. We even find C-3PO endearing. As androids get more human-like they tend to produce reactions of unease.

Here a near miss in terms of appearance and behavior translates into a big miss in emotional terms. The same built-in emotional bias seems to characterize our response to human-like robots.

From Romantic to Work Relationships

There are many differences between romantic relationships and relationships that occur in work contexts. The relationships we enter into with doctors who prescribe our antibiotics or baristas who make our cappuccinos tend to be more temporary. In the case of the barista who makes a morning espresso, they may be positively fleeting. But they share a feature of our romantic relationships. We have an interest in the mental lives of our doctors and baristas. Our social needs enmesh us in a very wide variety of relationships with other human beings. The relationships we enter into with our lovers or children matter a great deal to us. When one of those relationships goes badly it has the power to wreck your life. The relationships we enter into in work contexts tend not to have this power over how your life goes. But they still matter. When your barista pointedly ignores your morning salutation, your day goes a little worse. Our very social evolutionary history makes sense of this reaction. John Cacioppo and William Patrick call humans "obligatorily gregarious."[7] They say, "As an obligatorily gregarious species, we humans have a need not just to belong in an abstract sense but to actually get together."[8] The paradigm of a good life for an obligatorily gregarious member of the species *Homo sapiens* contains many positive social experiences—greetings offered and reciprocated, inquiries after your thoughts when you appear to be particularly pensive, offers of assistance with a difficult door, and so on. Someone who suffers from social isolation probably has fewer and poorer quality relationships of all types. We can understand the significance of these relationships to us if we view them as relationships between beings with conscious minds. Beings who share your most valued feature—they have minds—care about you. Other beings like you loathe you—but at least you matter in some way to them. The editing out of humans from our daily lives may lead to your needs being met in ways that are objectively superior. But it deletes these distinctively human aspects.

A 2015 study by researchers at Oxford University and Deloitte presents "waiter or waitress" as "quite likely" to be automated within two decades—a

90 percent probability. If we focus exclusively on efficiency, then this claim makes sense. An appropriate measure of efficiency for waiters takes into account accuracy in the taking of orders, the speed with which those orders are conveyed to the kitchen, the time it takes for ordered food to be sent to the right diners, the rate at which plates are dropped, the accuracy in totaling bills, and so on. It's not hard to imagine machine waiters being more efficient than their human coworkers. But we value the fact that our waiters are beings who are capable of feelings like ours. We have an interest in their mental lives. We like waiters who ask "Did you enjoy the goulash?" not because their programming reflects a finding that diners like to be asked such questions but because they have some genuine, if fleeting, interest in how you liked it. The value of humanness is likely to be entirely opaque to visiting extraterrestrials. They are unlikely to care about the difference between being served food by a machine or a human. Perhaps our best philosophical arguments in favor of the phenomenal consciousness of the waiter will seem absurd to them. But we humans find these arguments plausible, even if not fully philosophically convincing, and hence the difference matters to us.

Other jobs that center on human relationships are teacher, nurse, counsellor, actor, and writer. I've suggested that part of the value that we place on the performance of these jobs is the fact that they place us directly into contact with other humans. We receive the help that we need from beings who have minds like ours.

How does this suggestion that we value interacting with beings who have minds like ours support a Digital Age with human workers? Work is a key part of the solution to the problem of how we build a successful society out of humans who are total strangers to each other. Pre-Neolithic forager bands were face-to-face communities. Foragers typically treat strangers with suspicion. The economist Paul Seabright presents successful societies of strangers as one of our species' signal achievements.[9] He reflects on the improbability of such societies given our evolution as a "shy, murderous ape that had avoided strangers throughout its evolutionary history."[10] We are "now living, working, and moving among complete strangers in their millions." What makes this remarkable, according to Seabright, is that "Nothing in our species' biological evolution has shown us to have any talent or taste for dealing with strangers."[11]

Seabright gives much of the credit for this conversion of a shy, murderous ape into a gregarious, trusting human to markets and the institutions that form around them. Foragers must cooperate with members of their band—relatives or at least individuals they know well. Human societies thrive because of mutually beneficial exchanges. When hunters work together they achieve more than all of them could have on solo expeditions. Work translates these one-off mutually beneficial exchanges into long-standing arrangements. The cooperative enterprises of the technologically advanced societies of the early twenty-first century are significantly more complex than the group hunting or gathering expeditions attempted by foragers. Think of all the diverse contributions of individuals, many of whom are strangers, that go into supplying a house with power and ensuring that you can use that power to conduct a Google search. Jobs tend to standardize contributions to cooperative undertakings. When you become your village's blacksmith you advertise yourself as being the person to go to for objects made from wrought iron. You stand ready to accept the trade goods or money of total strangers in exchange for your handiwork. If you take a job as a software engineer working in Google's search division, you stand ready to meet the Internet search needs of many millions of total strangers. Work generates valuable goods. But it is also an important social glue that binds suspicious strangers into successful cooperating communities.

Humans may be obligatorily gregarious, but when left to our own devices that gregariousness tends to be parochial. We seek out people we know or people who resemble us in ways that we care about. We don't seek out strangers. In one study that demonstrated our preference for people like us, Angela Bahns, Kate Pickett, and Christian Crandall compare the social groupings that formed at a large state university where there would be lots of choice about who to bond with socially with the groupings of small colleges from the same geographical region.[12] They found that when given the choice, students used the greater choice of the large university to find others similar to them. They may say that they arrive on the very diverse campus of the large university excited about the array of different kinds of people they could form social bonds with. But this doesn't seem to describe how they behave once there. This preference for those we know or those who resemble us in some way that we judge significant makes sense in the light of human forager origins. For foragers, strangers are scary.

Work is a way to create bonds between strangers in large, diverse societies. It requires you to reach out beyond your forager comfort zone and form relationships with scary strangers. The fastest way for your coffee business to go under is to limit its services to friends and relatives. You want to serve coffee to all comers. A robotic barista may be a more efficient provider of lattes than its human equivalent. But when you receive your latte from the human you generate some of the social glue that fashions diverse strangers into harmonious societies. A brief visit to a Starbucks is likely to require social contact with someone your forager shyness suggests you should avoid.

Suppose you have the misfortune to be raised by racist parents. You go out into the world and find that you share a work place with members of the group you were raised to hate. The relationships you form with coworkers may be a powerful way to reverse this unfortunate aspect of your upbringing. There is reason to believe that working together to achieve a shared goal is an especially effective way to overcome mistrust.[13] You may have been raised to hate Muslims but when your employer places you in a group that contains Muslims and requires you to work with them to achieve a difficult goal, then the prejudice of your childhood faces a serious challenge. Cooperative relationships are not unique to work. If you sign up to play soccer on the weekend you may find members of mistrusted groups in your team. But we should acknowledge that work is an important venue for them. .

Those who value the diversity of the modern multi-ethnic societies and who accept or even express enthusiasm about the end of work should suggest some alternative source for the social glue supplied by work. But they must do more than merely imagine it. Perhaps it is possible to create a socialist paradise in which everyone joyfully complies with the proposal popular in nineteenth-century socialist circles and popularized by Marx "From each according to his ability, to each according to his needs." In this socialist paradise, it won't matter than some of the needy speak languages that differ from the majority, worship differently, and look different. A socialist paradise would be wonderful. We might set it as a long-term goal. In the meantime, however, we can use the institution of work to both give people meaningful lives in the Digital Age and ensure that they form bonds with strangers. To put it pithily, work works. It's something that we have. It's worth preserving it until we have a proven replacement.

How will we stop our forager parochialism from reasserting itself and restricting ourselves to friendships with people we judge to be like ourselves? I return to this issue in chapter 7 where I discuss the enthusiasm of advocates of the universal basic income for a world either without work, or at least with much less work.

The recognition of humans as obligatorily gregarious allows that sometimes we just want to be alone. A general preference for being with and dealing with other humans is a legacy of our evolutionary past. This general preference does not mean that we must always be with beings capable of feelings like ours.

Much is written about advances in medical robotics. I've suggested that, in general, we enjoy the benefits brought by human doctors and nurses. But sometimes in medicine we do want to be away from other humans. Some procedures are embarrassing. If a robot performs my prostate exam I have no grounds for awkward feelings about the state of my bum and worries about my decision to order the extra-spicy vindaloo for lunch. Here the absence of mental states is an advantage. But these cases should be considered exceptions to the general rule about a preference for interactions with other human minds. We are social creatures even if we sometimes want to hide out in our man or woman caves. This preference may extend to the seemingly very human activity of therapy. Writing in the *Atlantic Monthly*, Derek Thompson observes that "some research suggests that people are more honest in therapy sessions when they believe they are confessing their troubles to a computer, because a machine can't pass moral judgment." Thompson does not think that this means that therapists will soon suffer the fate of handloom weavers in the Industrial Revolution. "Rather, it shows how easily computers can encroach on areas previously considered 'for humans only.'"[14] The view of the Digital Age defended in this book presents this preference of machines as a marginal phenomenon—cases in which we take time out from other humans rather than reflecting a desire to restructure our lives to eliminate human interactions. I've suggested that we have a general preference for human romantic partners. But sometimes you'd prefer a few minutes with some sex tech to the *sturm und drang* of "making love!" with another human being.

There's another side to our interest in the conscious experiences of workers in the Digital Age. When we socially enhance the job of sales assistant we make it more enjoyable. Mihaly Csikszentmihalyi and Judith LeFevre's paradox of work suggests that our notion that leisure time is more enjoyable than work time is for the most part false. Some of the work in today's technologically advanced societies is a dull grind. But some of it is meaningful. If we choose to live in a society in which this work is automated, then we choose to eradicate many of these pleasurable experiences. I suspect that there is an inconsistency. We well understand the pleasure that we get from performing purposive activity such as taking an evening stroll or signing up for pottery class. People who enjoy strolls and pottery don't claim to be the most efficient at those activities. You would scoff at the suggestion that these and other activities that we enjoy be automated. The pleasure your inferior performances give you makes these activities worth doing. There's something enjoyable about getting clay on your hands. You should spare some regard for the positive experiences of the human nurse who checks your blood pressure and barista who produces your special coffee together with the signature design on its froth.

We can imagine that when the robo-psychotherapists of the very distant future tell patients that they know what it feels like to go through bereavement, they will mean it. I've claimed that we have a bias in favor of human experiences, but nothing I have said suggests that it's impossible that robo-psychotherapists could have these feelings. But this would be an odd and self-abnegating direction for us to take digital technologies. It's clear where there's a need for automation. A fully automated cockpit that gets me from Wellington to San Francisco more safely than a human pilot would be a digital technology worth having. But what could be true for pilots need not be true for roles based on our social natures. Why automate jobs that we are both good at doing and find deeply meaningful? A collective decision to automate these jobs is a bit like seeking to automate your evening stroll.

What Counts as a Social Job?

It may be relatively easy to see how barista and teacher are social jobs. But the suggestion that the central feature of such jobs is the making of connections between beings with similar minds enables us to see many other jobs as essentially social.

When you read Harry Potter you are reading something written by J. K. Rowling. Rowling's writing is a social activity that places her into contact with millions of readers. When readers speculate about Hogwarts they seek to gain some insight into her mental life. Rowling had distinctive thoughts and feelings when writing about the villainous Lord Voldemort. We are impressed that the entire Harry Potter universe was brought into existence by that very powerful human imagination. The revelation that all the Harry Potter books were written by a story-writing AI would not be merely interesting. It should be a profound disappointment to the books' many millions of fans—they have learned that there was no distinctively human consciousness on the other side of those pages.

The relationship between Rowling and her readers is asymmetrical. Rowling affects her readers in ways that they mostly cannot affect her. She writes the books and they read them. For certain kinds of human relationships some approximation to symmetry is important. In general, we want our romantic relationships to be symmetrical. We tolerate a significant degree of asymmetry in relationships that make lesser contributions to our well-being.

The point is not that humans cannot be tricked by machines programmed to write fantasy novels. The point is how we respond to such trickery. In chapter 2 I argued that our judgments about mind go deeper than the snap judgments of our Hyperactive Agency Detectors. Many of today's machines are granted provisional entry to the human chapter of the mind club. Our assessments of mind draw on more information than generated by the 5 or 25 minutes of conversational probing available to Loebner Prize judges. Assessments of mind are ongoing. When you learn that a chatbot has used the sexy-talk strategy to influence your decision about whether it was a member of the human chapter of the mind club, you seriously consider reversing your initial admission of it.

In their 2017 book *Machine, Platform, Crowd,* Andrew McAfee and Erik Brynjolfsson describe a chemistry professor and music aficionado who hears a work composed by the music-writing AI Emily Howell and pronounces it "one of the most moving experiences of his musical life."[15] The professor later hears a recording of the same music at a talk given by Emily Howell's programmer and says "You know, that's pretty music, but I could tell absolutely, immediately that it was computer-composed. There's no heart or soul or depth to the piece."

This reaction is not the absurdity that McAfee and Brynjolfsson present it as. The claim that the aficionado could "tell absolutely, immediately that it was computer-composed" is clearly false. But his assessment that it has "no heart or soul or depth" can be true so long as we understand it as drawing on beliefs about the mental states that lie behind its composition. Just as we can reverse our provisional inductions of Loebner Prize winners into the human chapter of the mind club when we learn more about them, so too we can reassess the value that we initially place on Emily Howell's compositions. We can view being moved by its compositions as fraudulent. Tchaikovsky wrote his *1812 Overture* with its volleys of cannon fire to celebrate Russia's defense against Napoleon's army. Any cannon volleys written into Emily Howell's compositions will have no connection to beliefs about heroic repulses of invaders. It's perfectly appropriate to grant that its compositions may count as "pretty music" but to retract any provisional attributions of "heart or soul or depth" when we learn more about how they were composed.

Can I Justify My Pro-Human Bias?

I suggested that given the choice between a human lover and a behaviorally identical robot driven by a digital computer, if you care about the mental life of your prospective romantic partner, you should prefer the human. I extrapolated a preference for human baristas and teachers from this preference for human romantic partners. Even the best arguments for the existence of human mental and emotional states in the ingeniously programmed sex robots of the future should not allay the suspicion that there are no human feelings behind their very human-like behavior.

Effectively, I am arguing for a pro-human bias. We prefer to interact with humans both in the bedroom and in the work place. The word "bias" may sound ominous. A bias in favor of working with humans seems to suggest a future quite different from that depicted in *Star Trek*. The crew of the Starship *Enterprise* combines a motley assortment of humans, half-humans, beings from other planets, and artificial beings. The first *Star Trek* series, filmed in the 1960s, sent a salutary message for a world beset by racial strife. It was a world of harmony between beings whose differences were objectively greater than those that seemed to challenge the America of the 1960s. In the 1990s remake *Star Trek: The Next Generation*, one of the crew is a cybernetic being—Data. The central message of this chapter would seem to justify rejecting Data. It would not have been edifying to hear Captain Jean-Luc Picard say: "I'm sorry Mr. Data, but I will not tolerate artificial crewmembers on my bridge!"

In what follows I offer philosophical defense of the pro-human bias. First, we should clarify the focus of this bias. Historical examples of bias involve treating some kinds of human as if they had a moral status inferior to other kinds of human. Slavery was purportedly justified because slaves had a moral status inferior to that of their owners. The moral relationship between slave owners and slaves was wrongly thought analogous to that between farmers and their livestock. Both slaves and livestock were property. Nothing analogous follows from the case for pro-human bias offered in this book.

I've suggested that it is rational for us to doubt that beings like Data have mental lives like ours. Rational doubt about the reality of Data's mental life may lead you to not date him. But it does not justify treating him as if he is a morally inferior being. Data is quite possibly something more than just a machine. He may be a member of the mind club. A rational recognition of the fact that he may have a mental life much like ours suggests that we should not treat him as we might treat an obsolete smartphone. Data is not property. He should not be simply recycled in the most environmentally friendly way when judged no longer useful.

My second point addresses the context of expressions of bias. How we assess any kind of bias depends a great deal on facts about the world in which it is expressed. There is, morally, a difference between expressions of bias which have actual victims and expressions of bias whose victims are merely counterfactual. Call the former *actual bias* and the latter *merely counterfactual bias*.

Actual bias has victims who can suffer significant harms. Societies should strive to eliminate biases in terms of race, gender, sexual orientation, religious creed, and so on. But we can think differently about merely counterfactual bias. This is victimless. A law that prevents Data and artificial beings from being employed as baristas or prevents artificial beings from marrying is victimless at a time when beings like Data do not exist and we lack the capacity to make them. It will be victimless at a time when we have the capacity to make beings like them, but have chosen not to.

Some racists seek to give their attitudes a positive spin. They complain that it is wrong to view them as opposed to members of racial group *A*. They are in favor of members of racial group *B*. Parents are allowed, and are indeed expected, to favor their own children. According to racism apologists we can preferentially benefit members of our own racial group in the way that we justifiably prioritize our own kin. We should reject these rationalizations of racism because of the real harm caused to a disfavored group. When members of the dominant group insist on hiring "their own" they may think no explicitly negative thoughts about the members of groups they overlook, but this pattern of preference is, nevertheless, harmful. Those from disfavored groups suffer unjustified disadvantage even if they are not actively loathed.

Now, consider these arguments when applied to merely counterfactual bias. If the members of a disfavored group are merely counterfactual they suffer no harm. We can express a preference for humans without worrying about whether Data or other human-like robots suffer.

Our moral assessment of expressions of bias depends on the context in which they are expressed. It's possible to imagine a future in which we have chosen to create beings like Data. That decision translates the counterfactual refusal to admit Data to Star Fleet or to hire him as a barista into a bias against actual beings. In this future society, this book may rightly be viewed much in the way that reasonable people today view Adolf Hitler's hate-filled *Mein Kampf*. My point is that we should think differently about bias with actual victims than we do about bias with merely counterfactual victims. No one need be harmed by a preference for human baristas if we don't create sentient artificial beings capable both of desiring employment in that line of work and of being harmed by a refusal to hire them.

Here is a philosophical thought experiment that demonstrates the acceptability of merely counterfactual bias. Merely counterfactual bias is a central feature of much of our popular culture. We fear the unknown and makers of movies exploit this by making the unknown seem as malignant and loathsome as possible. Perhaps supremely rational beings might reject all forms of bias, both actual or merely counterfactual. But we are not them.

Suppose that peace-loving extraterrestrials arrive on Earth. They offer friendship. They would like to live among us. They are not slyly seeking human extinction. By curious happenstance these new arrivals closely resemble the aliens of Ridley Scott's *Alien* movies. Their behavior does not resemble that of Scott's alien in the slightest. They use their telescoping jaws to consume a wide range of vegetarian delicacies. Their distant ancestors used the jaws to rip sentient prey apart, but those behaviors are as relevant to them now as are our ancestors' tree-climbing lifestyles to us now. Their respect for sentient beings leads them to scrupulously avoid injuries that would spill their acid blood.

I have described circumstances in which we should regret and apologize for our species' production and enjoyment of the *Alien* movies. The fear and loathing encouraged and exploited by the movies now have victims. In these circumstances, we should view the *Alien* movies as we now view movies in which all the villains have dark skin or cartoonishly Semitic features. If we insist on continuing to view the *Alien* movies, we should re-pixelate their villains. But what is true in this counterfactual scenario does not mean that we must cease watching Scott's *Alien* movies now. Until the arrival of the peace-loving extraterrestrials we go on fearing and loathing the alien. The victims of the bias inspired and exaggerated by the movies are merely counterfactual.

There are other merely counterfactual victims of bias promulgated by the *Alien* movies. Ash, who features in *Alien,* and David, who features in *Prometheus* and *Alien Covenant,* are evil androids. They gain our trust only to betray us. If we create sentient artificial intelligences, we should reflect on the hostility toward them generated by these movies. For now, however, no one need suffer due to any hostility toward artificial beings.

It can be useful to direct fear and hate at purely counterfactual targets. During the 1985 summit in Geneva between cold war adversaries the United States and the Soviet Union, President Gorbachev of the Soviet Union reported a conversation he had had with his US counterpart Ronald Reagan. Reagan had asked Gorbachev, "What would you do if the United States were suddenly attacked by someone from outer space? Would you help us?" Gorbachev reported that he had assured Reagan that the Soviet Union would come to the assistance of the United States. Reagan had returned the commitment.[16] Militarized antagonism directed against merely theoretical alien invaders was not directed against the actual civilian populations of the United States and the Soviet Union. Nuclear weapons placed into orbit and pointed outwards are not aimed at Moscow or Washington, DC. Bias directed at merely counterfactual beings is similarly victimless.

To future sentient machines who find these paragraphs contemptuous of their legitimate interests I apologize, but neither they nor those interests existed at the time of writing. I owe them an apology just as Ridley Scott owes an apology to future extraterrestrial visitors who just happen to resemble the aliens his movies have encouraged us to hate and fear.

It is entirely up to us whether we live in a world that contains beings like Data. In chapter 2 I suggested that the dominant motivation for AI is not currently directed at creating beings like Data. Rather it is pragmatic, focusing on the potential for machine learning to solve serious problems and generate wealth. We can derive immense benefits from AI without creating beings like Data. When we refrain from creating beings like Data we are not rejecting research on artificial intelligence. Rather, we are expressing a legitimate preference for the varieties of AI research that bring humans the greatest benefits. We should expect greater benefits to come from the pragmatic rather than from the philosophical focus. Maintaining this focus is a way to ensure that a merely counterfactual bias against sentient AIs never becomes actual bias against them.

Concluding Comments

This chapter has explored the value of humanness that I locate at the center of the social economy. The basis for our preference for human baristas, nurses, actors, and teachers is essentially the same as the basis of our preference for human lovers. We care about our connections with other beings who have distinctively human mental lives. In the possible future in which we create possibly sentient robots who express a craving for affection from other sentient beings we may have to rethink this. But we are not there yet. We can choose a human-centered future in which machine learners solve some of our most tractable problems without aspiring to sentience. In the chapter that follows I describe the contours of a social economy centered on human feelings and experiences.

6 Features of the Social Economy in the Digital Age

Societies of the Digital Age should center on two distinct economies defined by the values that they exemplify. The digital economy will center on the value of efficiency. It gives priority to means that better produce valued outcomes. Advances in digital technologies are ending the era of human supremacy in the domain of efficiency. We are no longer the world's best chess players. Soon we won't be the world's best accountants or truck drivers. The Digital Revolution will redirect human workers out of many vocations, but it need not end human work. Truck drivers can use their redundancy payments to reinvent themselves as social workers. The guiding value of the social economy is humanness. We care about efficiency in the social economy, but we often permit trade-offs that reduce efficiency in order to increase human participation and expression. You value the fleeting pleasures you get from a brief interaction with the barista who makes your coffee. It may not matter so much that, on rare occasions, she puts cow's milk in your coffee instead of the requested soy. We remember days ruined by unpleasant interactions with baristas and waiters.

This chapter explores differences between the goods of the digital economy and those of the social economy. One of the great challenges of the Digital Age will be to demarcate work that should be passed over to machines and work that should be reserved for humans. The boundary between the social and the digital economies is not fixed. Tasks do not present coded "digital" or "social." Rather, we must collectively reflect on what we value about a job. The decision that a job belongs in the social economy is founded on the value we place on interaction with other human minds. The decision that a job belongs in the digital economy draws support from a perceived dispensability of such interactions. As things are now, humans may be the most efficient means of achieving the ends that define these jobs, but reflection reveals that we don't lose much from the elimination of human minds. We want the mind work done but we don't care if it's not done by beings with minds.

If we decide that a job belongs in the digital economy, then we should seek reduced human contributions to increase efficiency. We should displace remaining humans with more efficient digital technologies. The defining value of the social economy is humanness. We value other humans in our personal lives, but also in our working lives. Efficiency matters here too, but we are often willing to accept reduced efficiency in exchange for more meaningful human interaction. Where declines in efficiency are a concern, we supplement human workers with technologies that correct errors. But we view it as imperative to retain human contributions. When we listen to a Yo Yo Ma recital we judge his contributions to be more important than the indispensable contributions of Petunia—his primary performance cello. We should make the same assessments of other digitally assisted social jobs. We should often look to socially enhance these jobs by increasing human contributions and making them more meaningful. Some roles present both the options of being made more efficient by the addition of digital technologies or being socially enhanced by amplifying human contributions. A single job in the pre-Digital Age could give rise to a very efficiently performed digital economy version and a resolutely human social economy version.

Two Economies for the Digital Age

The social-digital economy comprises two streams of economic activity centered on different values. The value that characterizes the digital economy is efficiency. The value that characterizes the social economy is humanness. These goods of the digital and social economies have different properties. In deciding where to locate a role we must decide what we value about it. Some jobs involve no or minimal interaction with human beings. It seems appropriate to locate these in the digital economy. We will want to choose the most efficient means of production. Human social talents are wasted here. Humans who perform these jobs should expect replacement by more efficient machines.

When we think about social jobs, interaction with other human beings is central. The social workers of the Digital Age will be aided by powerful digital technologies. But when we award credit and remuneration for the performance of jobs in the social economy we rightly grant these essential technological contributions a lesser significance. Today's poets need electric illumination if they are to keep writing into the evening. But they reject suggestions that electricity is a defining contributor to how they make their livings. The writing of poetry today is, in essence, no different from the way it was for the ancient Roman poet Ovid.

The drive for greater efficiency can be a spoiler in the social economy. It tends often to purge the joy from social jobs. Many teachers and nurses enter their professions in their twenties with excitement about the contributions they can make to children and patients. Attempts to increase the numbers of students taught and patients treated leave them and other social workers emotionally depleted. Teachers find themselves burned out by their forties and seeking new careers in real estate.

We should beware of illusory savings achieved by applying measures well suited to the goods of the digital economy, but ill-suited to the social economy. We do not adequately account for the goods of the social economy by counting customers served and medicines distributed. Applying the rules of efficiency to the social economy is as misguided as applying the rules of social economy to digital machines. If you are looking for empathy from your car's GPS when you get lost, then you are making a mistake. When you respond to your Fitbit's command to get up and take 250 steps by sinking deeper into your couch, your hyperactive agency detector may lead you to credit the device with feelings of frustration and disappointment. But there are no such things. Participants in the social economy enjoy interacting with beings who genuinely experience these feelings.

The suggestion of a social-digital economy has the disadvantage of going against an established trend in respect of work in the Digital Revolution. Many workers who deal most directly with human beings face replacement by digital technologies. People who work in customer service find their work automated. Machines are taking orders for food and serving it. This trend should be reversed. The first task of this chapter is to describe some of the features of the goods of the social economy and to contrast them with the goods of the digital economy. The Digital Revolution has ambiguous effects on work done by humans. Digital enhancement will tend to replace work done by humans with work done by machines. Social enhancement can expand the roles of humans in the economies of the Digital Age.

I should be clear about what is being claimed here. I am identifying humanness as something that we value in addition to efficiency. No comparative claim is being made. I am certainly not saying that humanness is more important than efficiency. If you're sick, the main thing you want from a medical professional is an effective therapy. If given a choice between a machine that spontaneously emits a cure for your cancer and an impressively empathetic human doctor who lovingly writes out a prescription for some homeopathic remedies, then you should choose the former. But that is to oversimplify the choice. The fact that efficiency matters more here does not show that humanness does not matter at all.

Suppose that we decide that, these points notwithstanding, we believe that some roles in Digital Age medicine belong in the social domain. When we learn of the inefficiencies of human medics—including misdiagnoses and prescription errors—we don't sigh and accept this as part of the price of dealing with doctors who have mental lives like ours. But we don't respond by passing all their tasks over to machines. To the extent that we value the humanness of our medical professionals, we seek to preserve the human contribution while seeking to use technology to remedy human inefficiencies.

Perhaps the human element of Digital Age medicine means that no matter how powerful are the technologies that pick up and correct the errors of human doctors, human error will always involve a rate of misdiagnosis that exceeds that of a purely digital counterpart. Whether we accept this miniscule increase in risk depends on how much we value the human contribution. We prefer to deal with beings who have minds like ours. We value it to the point that we are prepared to pay for it. There is an analogy between the goods produced by the social economy and those offered by the Fairtrade movement. "Fairtrade advocates for better working conditions and improved terms of trade for farmers and workers in developing countries."[1] The global success of the Fairtrade movement has shown that we are prepared, to some extent, to put our money where our moral-platitude-expressing mouths are. Those who pay for Fairtrade chocolate want the actual chocolate. When they hand over their cash they aren't satisfied with an expression of gratitude for the contribution to the cause unless it is accompanied by actual chocolate. They accept that they should pay a bit more so poor people who harvest cocoa beans receive more than their disadvantaged position in the global market suggests that they otherwise would. This is not straightforward charity. I suggest that we are prepared to pay more to preserve human contributors. We choose to spend more money on things done by humans when presented with the alternative of having those things done more cheaply by machines. We build technologies that compensate for the cognitive glitches and oversights of human teachers and doctors.

Some Noteworthy Differences between Social and Digital Goods

Costs

Jeremy Rifkin makes much of the zero marginal costs of the digital economy.[2] It costs a great deal to produce the first copy of a word processing software package. But once made it can be copied and spread at a cost that, if not zero, is close to it. The millionth copy of Microsoft Office works just as well as the first. The efforts of companies in the digital economy can scale. This propensity to scale is a celebrated feature of digital platform businesses. Those who start these businesses spend a great deal of money setting up a website and an app for minimal financial returns.[3] Suppose a ride-sharing business recruits sufficiently many drivers and passengers to dominate the Wellington market. It can, without much additional cost, expand into the more lucrative Auckland market. The goods of the social economy do not scale in this way. They have high marginal costs. We see difficulties in trying to scale social goods both in those who produce these goods and those who receive them.

The low marginal cost of digital goods can lead to enormous profits. We saw this in chapter 3 with Antonio Garcia Martinez's suggestion that Facebook's approach to advertising is guided by the "billion times anything" principle.[4] The goods of the social economy tend, by contrast, to resist multiplication. The person who records the words "We at Acme Inc. truly value your call" may manage to feel real gratitude as she records it. Since that recording is digital Acme Inc. can produce many copies of it. But the magnitude of the emotional load does not get multiplied by the number of people who hear the message. Rather it dissipates to zero, or close to it. When you hear the message, you have no sense of a feeling human being responding in some, admittedly fleeting way, to your specific plight.

The need for contact with humans when dealing with a company owned and run by humans seems part of the explanation for the popularity of the strategy of repeatedly pressing "0" as soon as the automated message begins. You know that humans run the business. You have the feeling that these humans are in some way responsible for the problem you are now experiencing. You want to express your complaint to a human. When you are forced to deal with a machine you feel you are being brushed off. You know that when you get to talk to a human there is the possibility of eliciting some feeling of regret for the shortcomings of the product you purchased or service you received. This need for human contact should not be interpreted as a fear of technology. It reflects something more primal—a desire to connect with other humans. It just feels better to complain to a human. It can be nice sometimes to hear "I know how you must feel" from a stranger over a telephone line even if it's not immediately followed by a plan to rectify the problem. If that line is expressed with sincerity you can feel some confidence that your woes elicited a brief feeling of empathy.

A company that wishes to convey genuine sorrow for a product glitch must employ sufficiently many human beings to meet the expected emotional needs of large numbers of disgruntled customers. Expressions of sorrow have high marginal costs. Writing "sorry" on a website or tweeting a 140-character version of "sorry" are less effective ways of sending the message that the humans at the company recognize that they were at fault and owe effective expressions of regret to human customers. Effective expressions of regret require actual human beings who present themselves as employees of the company that produced the faulty product. It's true that a CEO who pays employees to express regret for a faulty product need not actually feel regret. But he does pay the high price of many human expressions of regret that should ideally present as authentic to those who receive them. The US military understands that when soldiers are killed, that information must be conveyed to families in ways that emphasize humanness at the cost of efficiency. Human beings must be sent to the homes even if that delays the sending of the information. There is rightly little interest in signing up all families of soldiers to an app that will inform them the instant a loved one's death is confirmed.

The goods of the social economy take time to produce. Their production cannot be rushed. Many companies offer to address your problems with online text exchanges. When you engage in these conversations you wonder whether there is a human being typing the responses. Automation achieves the end of conveying information but often detracts from the human significance of the conversation. Amazon's side of the text conversations you have with it contains many statements of the type "It was a pleasure helping you!" and "Thank you for getting in touch with Amazon." The pleasantries that we use to express emotions turn out to be the most predictable elements of conversations. They are, consequently, easy to program as short cuts. When you type "thanks," a keystroke can promptly send the reply "You are most welcome!" When the pleasantry arrives much faster than the time it would take to type it, you know that it results from a preprogrammed shortcut. Amazon's expression of gratitude is obviously ersatz. It is associated with none of the emotions human recipients of the message feel should accompany the typing of those words. This is not to say that your relationship with an Amazon employee responsible for dealing with a complaint is of utmost importance to you. Learning that Amazon's side of the conversation was entirely automated would not be like discovering that your lover's head was filled with silicon chips and circuit boards. But that doesn't prevent it from feeling a bit off.

We see some effects of an overemphasis on efficiency on the producers of social goods. Jobs that require some emotional engagement can be exhausting. Overworked nurses burn out. If you ask an exhausted nurse in his early 50s contemplating becoming an Uber driver "Is this how you expected nursing to be when you entered the profession?" you are quite likely to be told "No!" It is wrong to seek the kinds of efficiencies possible for the digital economy in the social job of nursing. Digital technologies make a difference here. But they do not change the tasks which we should recognize as part of the essence of nursing.

Current working conditions that overburden workers give them little opportunity to participate in the human relationships that should be essential features of their jobs. Today, companies make these roles redundant because they have too shallow an understanding of the benefits brought by direct human contact. Sustained social contact can be exhausting. It's a sad fact that many people drawn to the professions of teaching or nursing by an excitement about teaching or helping those in need find themselves so burdened that they enter their forties in a state of exhaustion. If they stay in their professions, they aspire to ascend bureaucratic hierarchies to be able to do less teaching or nursing, the activities that inspired their twenty-year-old selves. The social work of teachers and nurses should be acknowledged for its true worth. It is emotionally taxing to enter and sustain relationships that generate social goods. An economic system that places little value on these goods forces those who provide them to overtax themselves.

Ease of Quantification

There is a bias in our evaluation of benefits for society toward those that are easy to quantify. We can count the numbers of copies of digital products. The benefits of social goods are harder to quantify. That is because many of their effects occur inside human heads. The success of an apology for a faulty product given by an employee to a customer depends on how the apology is delivered. The differences between successful and unsuccessful apologies are difficult to measure. The result that is sought, a feeling by an aggrieved customer that her complaint has been understood and responded to, occurs inside a human head. You can easily count the number of roses you give to your beloved. What is less easy to quantify is the emotional impact of those gifts. Sometimes a single rose says what ten roses fail to say. But the emotional message is what you are aiming at. The messiness of these emotional outcomes means that they are ill suited for corporate balance sheets. We should resist a bias toward outcomes that are easy to measure and quantify and away from those that are difficult to measure and quantify. A preference for easily quantified goods with straightforward measures is, in effect, an unwarranted preference for the digital over the human.

Benefits

Loyalty is a valued feature of human relationships. It is also valued by companies that seek to sell us stuff. There's money in the prompting of emotions that have their proper place in interpersonal relationships. Consider business loyalty programmes. Companies encourage us to be loyal to them and to their products. If you are loyal to your spouse, then you reject offers from objectively superior potential partners. You cite your marriage vows. Businesses see profit in loyalty because they see in it a willingness to pay more for objectively inferior products. The business view of customer loyalty achieves its apotheosis in the pricing algorithms of online vendors.

Here loyalty can come with a price. A loyal customer has a proven track record of buying. She is predictably willing to pay more. A new visitor to the site has no established record and must be enticed with lower prices.

This exploitation of loyalty is very different from expressions of loyalty that are fit and proper in the interpersonal domain. It is analogous to a case in which you treat your spouse's past record of forgiving infidelities as encouragement to cheat more. If you are loyal to your spouse, then you can expect similar in return. Your spouse should reject offers from potential partners who may, in objective terms, be preferable to you. This reciprocity is not a feature of corporate loyalty programmes. Here it's less "loyalty brings its rewards" and more "loyalty gets penalized." Casinos benefit when loyal customers prefer their less generous slot machines to the more generous machines of competitors. But don't expect Caesars Entertainment Corporation to start measuring the success of its casino loyalty program in terms of how much money it awards to problem gamblers.

In 1970, Milton Friedman famously argued, "The Social Responsibility of Business Is to Increase Its Profits."[5] Friedman rejected the idea that businesses should promote social ends even if those ends seem just. When they do so corporate executives pretend to be public servants. Friedman granted that businesses see profit in encouraging us to think differently. He allowed that it could be in the interests of a business to "cloak" its actions in terms of social responsibility. There's money in products advertised as environmentally friendly. There's money in Google's former motto "Don't be evil." There's money in casinos feigning loyalty to gamblers.

It would be naïve to think that we can change these basic facts about the behavior of corporations in capitalist economies. But we can respond differently to businesses when they pay people to represent them. A human worker is a more appropriate recipient of loyalty than an algorithm. A salesperson who sells many products to you may benefit financially from your loyalty. If she is a very good salesperson she may use her humanity to, in Friedman's terms, "cloak" her real purpose, which is to get you go buy more. But she is at least capable of reciprocating expressions of loyalty from customers. When corporations dispense with human front people and seek to direct us toward the more efficient option of purchasing or complaining online, they make no pretense at humanity. There are restricted opportunities to appeal to individuals who understand the human meaning of loyalty.

The presence of human employees should reassure customers. It's harder for human employees to exploit loyalty in the way that algorithms do. Human employees have the same basic understanding as customers of what it really means to be loyal. Imagine: "I want you to entice people to become our customers with low prices. Once you have made your first sale you should bond with customers and speak the language of loyalty and commitment. To the degree that you judge that this language is working you should take the opportunity to raise prices. Reduce prices only when you suspect that the language of loyalty is wearing off." Human salespeople who successfully implement this strategy might be expected to test high for the trait of psychopathy. But it's a lucrative sales strategy for machines. The humans who program these strategies into computers stand at some distance from customers who receive their market-directed punishment for their loyalty.

Human employees can act badly. They can lie to and cheat customers. The scams of Enron Corporation are examples. But there is an upside to human employees. A human employee of a misbehaving corporation can become a whistle blower. Algorithms don't blow whistles on themselves. They comply with the Milton Friedman business ethic that directed their creation. If you are worried about corporate malfeasance in this digitally empowered age, then you should celebrate the presence of each human employee. Their Friedman-esque indoctrination may run deep, but there's always the hope that your moral appeals can penetrate it.

Consistency

Consistency is a noteworthy feature of the goods of the digital economy. The millionth copy of a software program should be the same as the first. Consistency is also an important feature of the social economy, but often we value departures from it.

The replacement of the handloom by the power loom is sometimes presented as a simple replacement of worse by better—a more efficient way to produce textiles replaced a less efficient way. But this is an oversimplification. There are differences between handloom- and power loom–produced textiles. There is no uniform way to value these differences. Power looms are better at producing a uniform weave. Handlooms produce variation in the weave that may reflect lapses of attention or deliberate interventions to make a repetitive task more interesting. For many of the purposes of the emerging industrial economy a uniform weave was an asset. An interest in the human aspects of weaving suggests a different approach to these inconsistencies. Writing about an Indian government initiative to promote the handloom weaving industry, Lalithaa Krishnan says of textiles produced by handlooms "Variations and uneven finish are part of the charm."[6] Inconsistencies suggest a human connection. You could look at an item of clothing, notice the variation in weave, and enjoy feeling some connection with the human being who produced it. Was the human weaver just sloppy or wanting to relieve tedium by making some changes? To say that for many purposes these inconsistencies have no economic value, and indeed reduce a product's economic value, should not be taken to deny that variations and uneven finish cannot sometimes be economically valuable. In the social economy, we value a degree of inconsistency. If you order a flat white coffee, you want what you are given to be similar enough to other flat whites you have consumed. But you value variation in the way it is presented to you and in the patterns in its milk froth.

The copies of digital economy goods tend to be consistent. You are unlikely to be happy to discover that your version of Microsoft Word always italicizes footnotes possibly because your copy's coder was sloppy or possibly because she held the view that material relegated to the footnotes sometimes gets insufficient attention. One copy of a specific digital good should not differ from another copy. The goods of the social economy tend to be consistent *up to a point*. When you go out to a favorite restaurant you want items on the menu to be as you remember them. But you don't want to have exactly the same dining experience. Sometimes you sit near the window. Other times you sit near the aquarium. Sometimes you get a waiter who's keen to banter. Other times not. Sometimes there's an intriguing special on the menu that the chef thought she would try out. Sometimes there isn't.

An overemphasis on consistency is often the first step in the preparation of a job for automation. When an employer devalues the distinctively human aspects of a job, it can be a sign that she is preparing to farewell her human staff. Discouraging the distinctively human aspects of jobs compares humans for competitions against machines that they cannot win. The McDonald's fast food chain prides itself on offering a uniform experience to diners the world over. This elimination of latitude for its human employees to interact differently with customers makes McDonald's interest in replacing human servers with self-service kiosks unsurprising.[7]

Spatial and Temporal Range

The range of the goods of the digital economy reaches as far as the reach of the Internet. If you have taken an enormously cute picture of your puppy, a button-press or two suffices to distribute it to Facebook friends scattered across the globe. The cuteness of the puppy picture will be as apparent to someone on the other side of the globe as it is to a colleague sitting in the next office. A new Facebook friend who stumbles across the photo a year after your original posting can enjoy the photo in the same way as someone who looks at it a second after you first post it. Many of the most important goods of the social economy have limited spatial and temporal ranges. It's good to use Skype or FaceTime to connect with a loved one. But the connection you get tends to fall short of that which you would have were they in the same room as you. There are valued features of face-to-face connection that these spatially and temporally distant interactions fail to replicate. The mediation of technology makes a Facebook friend different from a real friend. There is certainly overlap between the categories. But there are many Facebook friends who are not real friends. Your Facebook frenemies are probably just your enemies. There's a big difference between a Facebook friend's disposition to like pictures you post and a demonstrated willingness to make sacrifices that we have long recognized as part of friendship.[8]

Sherry Turkle points to the pernicious effects of technologies' invasion of the realm of human interaction. She focuses on the threat posed by digital technologies to relationships. Social-networking technologies promise more connectedness and enhanced intimacy. They instead leave us more isolated. We find ourselves, in the words of the title of one of Turkle's books, "alone together." According to Turkle, in the age of Facebook and Twitter "we find ways around conversation. We hide from each other even as we're constantly connected to each other."[9] Digital technologies turn out to be imperfect substitutes for face-to-face conversation, conversation that "is open-ended and spontaneous, conversation in which we play with ideas, in which we allow ourselves to be fully present and vulnerable."[10]

Turkle documents many ways in which digital communication is a poor substitute for face-to-face communication. It purports to offer what we get out of face-to-face communication in a more efficient and compact form. But it seems not to. Turkle notes the spoiling effect of digital technologies on these most prized of interactions. The omnipresence of an attention-hogging smartphone means we tend not to be fully present in our conversations. We become accustomed to "continuous partial attention" as we await the next message alert and stand ready to temporarily exit whatever conversation we find ourselves in. We are learning the hard way about the inadequacy of digital interactions as substitutes for the traditional ways of communicating or responding to romantic intent. Turkle suggests that we reclaim conversation as "a step toward reclaiming our most fundamental human values."

Turkle's principal focus is on the effects of the digital technologies on our most important relationships. We should see the same spoiling effects on relationships that occur in the context of work. The social goods of many of our multiple work relationships are best delivered face-to-face.

The Ambiguous Digital Futures of Sales Assistants

The distinction between jobs that belong in the social economy and those that belong in the digital economy seems straightforward. The former jobs involve interactions between human beings in which we place value on connections between human minds. In the latter, relationships with other humans are not essential. When we replace humans with more efficient digital technologies we come out ahead.

This makes our choice seem easier than it is. The tasks in Digital Age economies will not come coded "social" or "digital." Often there will be human contributions and digital contributions. We have two paths of improvement available to us. We can increase efficiency by introducing new or superior digital technologies. We can socially enhance by expanding the scope for human contributions. In some cases, we may see the simultaneous pursuit of both avenues of enhancement. The direct effects of digital technologies will eliminate some human workers. As human labor is freed up from parts of the economy best suited to digital technologies, it can be put to new uses in the social economy. In some cases, redundant workers might be re-employed in socially enhanced versions of their former jobs.

One of the jobs whose anticipated extinction few people seem to lament is that of sales assistant. We look to digital enhancements as sparing people from the mind-numbingly boring task of long days seated at cash registers endlessly scanning barcodes. The next step is the Amazon Go grocery store. It features "Just Walk Out" technology that permits customers to simply walk out with whatever they want without ever having to queue at a checkout.[11] Shoppers at Amazon Go download an app to their smartphones and present its customer-specific barcode to a scanner as they enter the store. Various sensors track the movements of shoppers, correlating them with the removal of items from the shelves. The store registers their departure and charges their account. This seems like an exciting possibility. Few supermarket shoppers will mourn the demise of checkout queues. But before we send sales assistant to the same vocational trash heap as the handloom weaver we should consider the possibility of socially enhancing sales assistance. There could be two Digital Age destinies for sales assistants. Some will be replaced by "Just Walk Out" technology. But others will find themselves liberated from their checkouts and empowered to become socially enhanced sales assistants.

An example of a socially enhanced sales assistant is a personal shopper. A personal shopper is someone who comes to know the person he or she assists and helps them to find products that are good matches for them. Rich people pay for personal shoppers to assist them to make the right purchases. They expect personalized service from people trying to sell them things. In a memorable depiction of high-end shopping in the 1990 movie *Pretty Woman*, a rich businessman played by Richard Gere takes his prostitute date, played by Julia Roberts, into a snooty Los Angeles boutique. Gere's character is clear about what he expects—"You know what we're gonna need here? We're going to need a few more people helping us out. I'll tell you why. We are going to be spending an obscene amount of money in here. So we're going to need a lot more help sucking up to us, 'cause that's what we really like." Fortunately, the sales assistant responds in the affirmative "Sir, if I may say so, you're in the right store, and the right city, for that matter!"[12] We can suppose that Gere's character would reject the suggestion that his and his date's tours of elite Rodeo Drive boutiques be made entirely nonsocial by the introduction of "Just Walk Out" technology. Don't expect the technologies behind Amazon Go to be coming to Balenciaga, Bulgari, Burberry, and the like anytime soon. The social side of the social-digital economy could create many opportunities for personal shoppers to provide these benefits to people whose means are more modest. A personal shopper would acquaint him- or herself with your preferences. He or she will combine this information with knowledge about a district's shopping opportunities. High fashion need not be the exclusive focus of the personal shoppers of the Digital Age. They might advise us on our culinary options. Personal shoppers with a culinary focus won't do this by implementing an algorithm that draws on large amounts of data about food and drink matchings but because they remember enjoying the combination of root beer and sushi.

In some cases, digital enhancement seems a much more likely outcome than social enhancement. Human pilots cannot hope to match the efficiencies offered by the fully automated cockpit of the future. The safety statistics of computer-controlled planes will predictably be superior to those of planes flown by humans. This does not settle the question of whether it is better to be flown by a machine or by a human. People frequently act in ways that increase their likelihood of death and feel right to do so. They go mountain biking on poorly marked hill tracks and then return home on heavily trafficked roads. For it to be worth incurring the additional risk of being piloted by a human, passengers need to get something out of interacting with that pilot. There were times when passengers had some direct experience of the human face of piloting. But the security measures since 9/11 have isolated pilots from passengers. Contact between pilots and passengers is less a social opportunity than a potential security breach. The days when a pilot might wander down the aisle asking "Which lucky child wants to come and help me fly the plane today?" are long gone. Those of us who are paying attention might cease watching a downloaded movie to listen to what the captain says over the intercom just after takeoff. If pilots had a long-term interest in protecting their jobs against automation, then they ought to have resisted some of the security measures that isolated them from passengers. When humans are decisively shunted out of the cockpit, we will barely notice their departure when they are replaced by more efficient autopilots. We accept the costs of humanness in some lines of work but not in others. We are more likely to forgive a misheard coffee order because our human barista was distracted than we are to forgive a fatal crash that occurred because a human pilot, distracted by a malfunctioning dial, failed to notice an abrupt descent in altitude.

The Different Digital Age Futures of Uber and Airbnb

Uber and Airbnb manifest a degree of divergence in approaches to digital and social enhancement.[13] The journalist Brad Stone presents both as helmed by extroverts with training in design who differ from the geeky introverts who built Google and Facebook. There are similarities between Uber and Airbnb. Both are San Francisco-based platform businesses that have, to use the preferred verb of digital entrepreneurs, "disrupted" entrenched incumbents, Uber the taxi industry and Airbnb the hotel industry. Airbnb became the world's most profitable hotel business without even having to acquire or build a hotel. As we saw in chapter 3, the principal value of a platform business is in its network. The successes of Uber and Airbnb come from two-sided markets that bring together people who need lifts or rooms and people prepared to accept payment for offering lifts or rooms. The interactions between passenger and driver or between host and guest happen on Uber's or Airbnb's platforms. They match sellers with buyers and take their cut. Both businesses understand the PR value of gestures of support for people whose experiences of these two-sided markets don't quite match the Uber or Airbnb ideals. But the whole idea of Uber and Airbnb as facilitators of two-sided markets is that people who choose to participate in these markets accept responsibility for any misfortunes they suffer. Anger and legal action should be directed at the offending parties on the other side of the exchange and not at the platform business which aspires to know as little as possible about the specifics of each transaction.

There are many similarities between Uber and Airbnb. But they may have different attitudes toward social work in the Digital Age. Chapter 1 discussed the emerging technology of driverless cars. Stone describes Travis Kalanick's excited reaction to Google's demonstration of a driverless car prototype: "The minute your car becomes real, I can take the dude out of the front seat … I call that margin expansion."[14] Uber's drivers protest the fact that it retains a fifth of the money paid by passengers. We've seen how the "billion times anything" principle turns small payments by Facebook's advertisers into a lot of money for Facebook. By July 2016, Uber had completed 2 billion rides.[15] Suppose it did not have to pay the dude in the front seat. Currently, Uber's drivers pocket 80 percent of the money paid by passengers. Two billion times 80 percent is an impressive expansion of Uber's margin.

Airbnb's reaction seems to be the opposite. In 2016, it enhanced its offerings by adding new "Experiences" and "Places" tabs to its website. Brad Stone describes Experiences as follows.

> With Experiences, travelers would be able to buy unique excursions, like hunting for truffles in Florence and visiting literary landmarks in Havana. Local entrepreneurs and celebrities would create and conduct these tours themselves.[16]

Airbnb Experiences would vary in price. Stone describes a US $800 Sumo experience with the retired champion Konishiki Yasokichi. It featured attendance at a training session, tournament seats, and a sumptuous meal. Airbnb Places would connect visitors with local hosts who would be able to direct them away from a narrow list of stereotyped tourist activities—think of the hordes of tourists who arrive in Rome or Paris and upon arrival can think of no alternative to joining massive lines for the Coliseum or the Louvre. Visitors connected with authentic Romans and Parisians would be able to spend their brief stay having experiences normally available only to those in the know. They might ask "Can you show me the 'real' Rome? I want to eat lunch where the Romans eat, not at a tourist trap within sight of the Coliseum." If they ask "Can you take me to the galleries that display the works of Paris's artistic *avant garde*?," a good host won't take them to the Louvre, a museum that she probably spent much of her adolescence visiting. Brian Chesky, the company co-founder, places emphasis on the social side of Airbnb's new offerings. He says "So can only people drive cars? I don't know. But only people can host other people. Only people can provide care. If you want something handmade, only a person can do that."[17]

Chesky certainly seems more alert to what humans can offer each other than does someone excited about getting the "dude out of the front seat." This does seem to be a vision of hospitality in the Digital Age in which humans direct their agency away from jobs that machines may do better than them—driving cars—and into areas that demand their social talents—enjoying helping a clueless stranger to experience the truly distinctive sights, tastes, sounds, and smells of your neighborhood. Chesky doesn't want to get rid of the dude who's showing you around Osaka.

If we are to fault Chesky it would be his jarringly technological presentation of these pleasures. Human experiences are being repackaged and marketed as Airbnb "Experiences." He says of this latest enhancement of Airbnb's services "I want to do for travel what Apple did to the phone."[18] For the time being, if you are about to fly to Madrid, a city that you do not know, Airbnb's website may be the best way to find someone to take you out to the city's best places for cider from Asturias. But when you return from Madrid Chesky wants you to grant Airbnb much of the credit for the fun times you had. I hope for a future in which we make use of the services offered by Chesky but feel less beholden to him and his inheritors.

As the shock and awe of a website capable of near-instantaneously connecting people who want to drink authentic Asturian cider in Madrid with strangers who would accept a modest fee to take them to the best cider joints wears off, we will be able to reassess the value of the contributions made to this experience. The Internet is today's wonder technology. We know from the case of electricity that yesterday's wonder technologies can become today's boring enablers. Today we can barely imagine the excitement that people must have felt when electricity brought artificial illumination to their living rooms. It's hard to think of places that you might visit that aren't artificially lit. You might investigate whether a hotel you are considering booking has a Wi-Fi router, but you just assume that, unless you've chosen a place that celebrates its remoteness, you'll be connected to the electrical grid. To say that electricity has become boring is not to say that it has been dispensable—quite the opposite. Our lives are more dependent on electricity than were the lives of people who viewed it as a wonder technology. Few of the digital technologies that are this book's focus would be possible without it. We expect cheap, reliable supplies of electricity and resent threats to this supply. We shouldn't forget the anger directed at Enron Corporation when it sought to profit by playing games with people's electrical supplies.[19]

The way we share out credit for a service has consequences for how we allocate its economic rewards. Today's platform businesses claim a significant share of these rewards. Uber retains a fifth of the money paid by passengers. It can because drivers and riders understand the power of its network. Airbnb does well out of its network. As soon as a guest makes a booking on the Airbnb site a "guest service fee" is displayed. This is charged until the host has confirmed the booking. According to the Airbnb site, "Guest service fees are typically 6–12% but can be higher or lower depending on the specifics of the reservation."[20] In addition to this, hosts are charged a "host service fee" of 3 percent calculated before the application of local taxes. I am not suggesting that we should intervene now to overturn these arrangements. Rather I am suggesting that a focus on the long view suggests a future in which people reject this way of giving credit for the services they receive. It would be hard to imagine today's cardiac surgeons tolerating an arrangement in which between 10 and 20 percent of the value of their procedures went to the company providing what they would view as platform *services* even if they were grateful for the additional business brought by the cardiac surgery platform business. We can hope for a future in which those who use the Internet, or whatever technology has replaced it, to offer their services to tourists will judge Airbnb's 6–12 percent plus 3 percent to be extortionate. We have no way to insist on these assessments today. Access to Airbnb is required to access the financial rewards and social pleasures of taking a foreigner to your favorite cafés. Currently Airbnb is indispensable. Perhaps the relationship between future social networking companies and people who enjoy Airbnb-style "Experiences" will be like those that obtain between the pleasures that we get from electricity and the companies that supply it. If a consortium of power companies were to credibly threaten to disconnect us from the grid many of us would probably consent to pay a lot more than we currently pay to avoid this eventuality. We look to elected government to prevent this eventuality. There is hope for socialized control and access to increasingly essential digital services.

We can expect people of the Digital Age to use the Internet or its successor to find strangers willing to show them around for a modest fee. But we should accept that almost all the value of these lies in the people who provide them. Today's poets need to pay their power bills if they are to keep writing into the evening. But they tend to dedicate the poetry collections to their loved ones rather than to the company that provided their power. A readjustment of attitude could make Internet services tomorrow's boring enabler. You pay for the services of a boring enabler, but not too much.

Space Exploration as Social Work

In some cases, we will need to think carefully about what we find valuable about a role. Consider the role of space explorer. We rightly celebrate the achievements of human space explorers. We venerate the first human in space—Yuri Gagarin—and the first human on the moon—Neil Armstrong. Currently, skilled humans are essential to space exploration. But the Digital Revolution should force us to think carefully about the value we place on human space exploration.

There is a way to understand what space exploration is about that emphasizes efficiency and suggests the imminent extinction of the human space explorer. This view presents the acquisition of data as the purpose of space exploration. The growth of our understanding of the universe and our place in it grows as we gather and analyze more data. We dispatch missions into space to gather data that cannot be procured by telescopes or other Earth-based technologies. We then analyze this data to reveal new truths about the Universe and our place in it.

This perspective finds the ongoing interest in sending humans into space or to other planets somewhat perplexing. Consider the interest in sending humans to Mars. The conditions humans will encounter during the flight to Mars and on the planet itself are very unlike those for which of minds and bodies evolved. There are doubts about the capacities of human psychologies and physiologies to withstand the voyage to and from Mars. Human participation comes with a very high cost for planners of space exploration missions. Much of the equipment that will need to be packed into a mission to Mars will have the purpose of keeping the humans alive and happy. Human astronauts will need the company of other humans to avoid going insane on the several months' trip to Mars. Each additional crewmember enlarges the technological infrastructure required to keep humans breathing, hydrated, fed, and happy. These hassles should be placed in a context that includes the many successes in sending uncrewed missions to Mars.

It's true that there have been mishaps that might have been corrected by human crewmembers. In 1999 NASA's Mars Climate Orbiter burned up in the planet's atmosphere as the result of a failure to convert nonmetric pounds into metric newtons. It's possible that a human crew could have corrected the error. But this expectation placed on that human crew member pushes faith in human exceptionalism too far. The inclusion of humans comes with costs. The failure of the Mars Climate Orbiter was a setback. It would have been a tragedy had a human crew been on board but failed to notice the error.

An interest in efficiency directs that mission planners accept the burden of sending humans to Mars when humans offer benefits that more than compensate for the handicaps they impose on the mission. Here's one suggestion about how we might work out when it's appropriate to send humans. "Robots … are very beneficial for missions that require precise and repetitive measurements or maneuvers, as well as missions that last a considerable amount of time. Humans, on the other hand, are much better suited to tasks involving decision-making, or those that require constant adjustments and intervention from scientists."[21] The value of human decision-making is well illustrated in the story of Apollo 13, a failed 1970 mission to the moon. Human ingenuity saved a craft that seemed doomed when there was an explosion in its oxygen tank, venting the indispensable gas into space. It's hard to imagine that NASA could have designed technologies with the general problem-solving capacity of Apollo 13's three human astronauts. They applied their flexible human minds to solve problems that could not have been anticipated by the humans who planned the mission.

We can be impressed by the heroics of the Apollo 13 astronauts. But we should also beware of the influences of the twin distortions of present bias about machines and the belief in human exceptionalism. We should expect exponential advances in the technologies that we send to Mars and use to gather data about it. Tomorrow's exploration technologies will not be handicapped by the limitations of today's machines. The same basic technologies that are piloting driverless cars and drones can easily be placed into space ships.

The making of Ron Howard's 1995 movie *Apollo 13* suggests a different, social interest in human space exploration. As the digital technologies that surround human space explorers improve we may seem sadly out of place. But we should acknowledge that that assessment is grounded in the value of efficiency. The inefficiencies of human space explorers will increasingly distinguish them from the technologies that surround them.

When we view space exploration as a social job then humans are the one truly indispensable element of the mission. When we hear Neil Armstrong's "One small step for man" line we are licensed to imaginatively place our feet into his oversized astronaut boots. We wonder what was going through Neil Armstrong's head when TV watchers on Earth heard him say "That's one small step for man, one giant leap for mankind." Did he mean to say "That's one small step for *a* man, one giant leap for mankind"? We vicariously experience the excitement of the first human being to step onto a different world. An estimated 600 million TV watchers got to vicariously experience stepping onto an alien world.[22] We imagined Armstrong's excitement and wondered whether the second man on the moon, Buzz Aldrin, experienced disappointment at his relegation. Some may be tempted to empathize with an abandoned Mars rover. After many years of heroic service, the *Spirit*, or MER-A (Mars Exploration Rover-A), lies abandoned on Mars. No one is contemplating the kinds of measures that brought the fictional human astronaut Mark Watney back to Earth in the novel and movie *The Martian*. It's tempting to try to empathize with *Spirit*. But it harbors no feelings for us to connect with. The rover comes as close to feelings of abandonment as a jettisoned toaster.

Human astronauts are important because our experiences are important. It's important for us to send humans to Mars so all the rest of us can vicariously go there. We get a vicarious pleasure from watching Garry Kasparov that we don't get from Deep Blue even if the latter plays better chess than the former. We want to know what it's like to be on Mars. When we send humans to Mars we make going to Mars social. We look forward to the explorers telling us about what it's like to walk the surface of the red planet. Those who present space exploration as a social job care about the inefficiencies of human space explorers. They seek technologies that rectify our inefficiencies and to accommodate our awkward needs. This focus is appropriate if we accept that humans are an indispensable part of the enterprise of space exploration.

When Sally Ride went up into space on board the space shuttle *Challenger* she became the first American woman in space. She performed her tasks on board the space shuttle *Challenger* with great efficiency. But she performed an additional social role. She went into space as a woman. Before she went into space, Ride confronted denigrating questions about whether space flight would damage her reproductive organs and whether she was inclined to weep when things went wrong.[23] Her step on board the space shuttle was a giant leap for womankind. We have many stories about young boys inspired by Neil Armstrong's first moon walk. Young girls could more easily identify with Sally Ride and vicariously travel into space.

Suppose we were to permit the value of efficiency to completely dominate space exploration. Developments of the digital package promise a future in which humans have no role to play. Probes will be launched into space. They will gather data some of which will be analyzed by on-board machine learners and some of which will be transmitted to Earth to be analyzed by more powerful machine intelligences. This is a future that doesn't differ greatly from Hollywood dystopias in which machines rebel against their human creators. There's some sense in which the machines will be exploring space on our behalf. But human choices will have little ongoing relevance to their explorations. The mission controls filled with jostling, sweating, chain-smoking humans depicted in iconic movies about the first

Apollo missions will have been replaced by machine intelligences that seek out patterns in incoming data with great efficiency.

Concluding Comments

Chapter 5 grounds the social economy in interactions between human minds. This chapter explores differences between the goods of the digital economy and the goods of the social economy. Many goods of the digital economy have zero marginal cost, or close to it. The businesses of the digital economy tend to scale. An app that well serves a town can scale to cover a continent. These properties tend not to be features of the social economy. The goods of the social economy are emotionally costly. It's easy to upload a film of an emotionally wrenching stage performance onto YouTube, but its emotional impact is significantly diminished. Seeing a YouTube clip of Chiwetel Ejiofor as Othello just isn't the same as being there. The efficiencies of the digital economy can transfer human labor out of areas for which it is ill suited. Some of today's jobs could face an ambiguous future. We could see digitally enhanced versions of them in which machines completely displace human workers. We could also see socially enhanced versions of them that grant greater scope and importance to human contributions.

Some of this might seem all well and good. But what chance is there that we will realize a social economy with a diverse collection of socially enhanced jobs? In the next chapter I explore how we should respond to this question. I suggest that we refrain from interpreting it as a prediction. We should instead interpret it as an ideal worth struggling for.

7 A Tempered Optimism about the Digital Age

I have described a way for humans to inhabit the Digital Age in which we benefit from fabulously powerful digital technologies that do not hog our attention, permitting us to focus more directly on the needs of other human beings. This could give rise to a social economy containing socially enhanced versions of many of today's jobs. The social economy does not get much attention in this time of hysteria about all things digital. As we become more accustomed to digital technologies they may cease to obsess us. Twitter and GPS-equipped smartphones could traverse the path already traveled by electricity. Yesterday's wonder technology predictably becomes today's boring enabler. The extreme wealth of today's champions of digital technology won't seem quite so right to those who enjoy its services. We will come to view them as we now view the 1900s robber barons of the energy industry and counter their hijacking of the digital economy.[1] This chapter offers a tempered optimism about the social digital economy. Not only is the social-digital economy an attractive vision of our collective digital future, it is achievable.

Part of the task of this chapter is to compare the social-digital economy with competing ideals for the Digital Age. I argue that its combination of attractiveness and practicality places the social-digital economy ahead of other ostensibly appealing visions of the Digital Age. I discuss two alternative visions. The first of these competing ideals is Jeremy Rifkin's proposal for a Collaborative Commons.[2] In Rifkin's vision of our digital future, people use digital technologies to create and share, disproving the bleak capitalist assumption that we must view new technologies as means of augmenting our own personal stores of wealth. I suggest that Rifkin offers, at best, a partial perspective on our collective digital future. Its principal appeal is to those with sufficient understanding of the digital package to use it creatively. A second seemingly attractive vision of humanity's digital future includes a universal basic income (UBI). In this view, our hopes for a flourishing humanity in the Digital Age should not reside in the creation of new varieties of work. Rather, we should celebrate our impending jobless future. According to its advocates, a UBI will permit the humans of the Digital Age to free themselves from wage slavery and to live as they choose. I respond to this by offering a defense of the work norm. According to this work norm, work is something most us should expect to do when we grow up. It is an expression of our social natures in which we make our livings by making some positive contribution to our societies. It facilitates the contacts with strangers essential to the flourishing of the complex and diverse societies of the early twenty-first century. We make some positive contribution to our society of strangers and thereby claim a portion of the wealth generated by that society. The work norm need not endorse the tedious and degrading conditions of many of today's jobs. We should reject the double standard that applies to the work of rich and poor people. The work of poor people seems described by the economists' term "disutility." Their work is not fun and it's not really supposed to be. The less well off turn up to work only because its disutility is compensated for by the positive utility of their pay packets. Many better-paid people speak truly when they say of their well-remunerated work that "it's not really about the money." They expect their work to be meaningful and enjoyable. They justify their high wages not by pointing the extreme personal disutility of what they do, but instead by pointing to the magnitude of their contributions to society. Removing this inconsistency would permit us to design social jobs that, although less well paid than the best-paid social work of movie stars, are still enjoyable. The work norm need not and should not support stigmatization of the jobless. There is a difference between work being normal and its being universal.

The Different Logic of Predictions and Ideals

The economist Paul Romer makes a helpful distinction between two types of optimism.[3] There is complacent optimism. Romer offers as an example of this kind of optimism "the feeling of a child waiting for presents." The more proactive conditional optimism "is the feeling of a child who is thinking about building a treehouse. 'If I get some wood and nails and persuade some other kids to help do the work, we can end up with something really cool.'" Romer suggests that we should be conditional optimists about technological progress. We shouldn't suppose that we can relax and expect the goods of improved technologies to just arrive. We won't get these unless we make the right choices.

In what follows I explore another way to probe the difference between proactive and complacent attitudes toward the future. There is a tendency for some advocates of a digital future to present their visions of humanity in the Digital Age as predictions. Consider, for example, Jeremy Rifkin's view of our collective digital future.[4] Rifkin celebrates the power of the Internet to connect people. The Internet brings individuals together to confront tyranny. Witness the role of Twitter and Facebook in facilitating the Arab Spring. Social networking technologies build on a fundamental human interest in working together to create value. The Internet is enabling the creation of what Rifkin calls a Collaborative Commons built on our need to connect and share. He says "While the capitalist market is based on self-interest and driven by material gain, the social Commons is motivated by collaborative interests and driven by a deep desire to connect with others and share."[5] Rifkin continues. "The result is that 'exchange value' in the marketplace is increasingly being replaced by 'shareable value' on the Collaborative Commons." Rifkin continues a line of thinking about the way we use and benefit from new technologies initiated by the futurist writer Alvin Toffler in the early 1980s. Toffler coined the term "prosumer" to describe someone involved in the production of what he or she consumes.[6] The Digital Revolution accelerates the prosumer movement. It transforms us from passive consumers viewed by corporations as dumping grounds for their overpriced digital products into empowered prosumers passionately engaged in the process of creating the products we use.

Is this an ideal or a prediction? It is important to note that the logic of an ideal conflicts with that of a prediction. When you present a view of the future as a prediction you present it as something that will happen. When you present a view of the future as an ideal you suggest that it requires effort on the part of your audience. Someone who predicts the sun's rising tomorrow suggests that it will happen regardless of what we do. No one approaches evening resolved to fight so that the sun may rise the next morning. William Wilberforce presented the abolition of slavery as an ideal that made significant demands of those living in a slave-holding Britain. He didn't suggest that those listening to him idly sit back, confident in his forecast of a slave-free Britain.

There's a gap between this advocacy of ideals and Romer's conditional optimism. You can present something as an ideal at the same time as expressing pessimism about its realization. For the record, I am a climate change pessimist. Climate change pessimism is not climate change denial. It concedes the reality of anthropogenic climate change. Moreover, it is compatible with the suggestion that it's possible for us to work together to reverse some of the effects of anthropogenic climate change. Climate change pessimists doubt that we will. The US election results of 2016 suggest that we will continue to focus more on waging wasteful wars on terror and building border walls than on arresting climate change. A leader who, under duress, utters the words "I believe in the reality of anthropogenic climate change" is unlikely to commit the resources to do much about it. Doing something about climate change requires more than the recognition that it is a real phenomenon with the potential for very bad consequences. The leader must prioritize preventing or mitigating these consequences over other more politically advantageous ends. If I'm asked to place a wager I'm betting that we will suffer the ill effects of climate change with the poorest among us expected to bear the heaviest burden. I hope to lose that wager. For that reason, I advocate the ideal of a truly green economy.

If the Collaborative Commons is a prediction, then it's something that will happen almost regardless of what we do. We created the Internet which, in turn, enables shareable value. And so, once we've traversed a few spasms of rejection by those who seek ownership over the Internet, we arrive at a Collaborative Commons.

There are two reasons to doubt the viability of the Collaborative Commons as a prediction about the Digital Age. First, and more generally, there are good grounds to doubt any prediction about the Digital Age. Technological packages are compatible with a variety of social arrangements. The Industrial package yielded Gilded Age America with its robber barons, the Soviet Union of the 1930s with its purges and deliberate mass starvations, and 1970s social democratic Sweden with its generous welfare safety nets. There are likely to be very many social permutations that can be made to fit the digital package. Rifkin's Collaborative Commons is one candidate. But there are no grounds to rule out, *a priori,* the various digital dystopias described in science fiction in which a few control all the machines and the rest of us have minimal access to them and are therefore excluded from the Collaborative Commons.

Rifkin's somewhat sunny vision of the Digital Age is no more inevitable than are more miserable interpretations of our collective experience of the Digital Age. Consider Marc Andreessen's separation of the humans of the Digital Age into two groups: "People who tell computers what to do, and people who are told by computers what to do."[7] The much-lamented rise in inequality might mean that we end up with something like this. The 1 percent will turn into the 0.1 percent who own the computers and therefore get to tell them what to do. The 99 percent will turn into the 99.9 percent who will be told by computers what to do. Andrew Keen forecasts a "winner-take-all, upstairs-downstairs kind of society."[8] He describes a world in which a near unbridgeable gap separates the many servants from the digitally enhanced few.

Current trends seem to make a digital dystopia described by Andreessen and Keen more probable than the Collaborative Commons or the social-digital economy. It seems to be the direction in which we are currently headed. In chapter 3, I challenged the idea that the putative desire on the part of information to be free will ensure that access is available to all. The sustainability of bound information depends on the resources that accrue to those who bind it. The fortunes of Google and Facebook buy many lawyers and politicians who will protect bound information even as their founders profess their commitment to the ideals expressed by Rifkin. Mark Zuckerberg, Sergey Brin, and Larry Page should be judged more by what they and their companies do than by the sweetness of their words.

I do not offer the social-digital economy as a prediction. There are many ways in which we can fail to realize a social-digital economy. When it comes to rosy forecasts of the future we should keep in mind Leo Tolstoy's dictum in *Anna Karenina* about happy families: "All happy families are alike; each unhappy family is unhappy in its own way." There may be more than one way for humans to achieve collective happiness in the Digital Age, but in the space of future possibilities, the digital dystopias do seem to outnumber the digital utopias. There are some bleak omens for those who support the idea of a future built around social-digital economies. Among the jobs most threatened by the current fad for economic austerity are those that should be in the greatest demand in a future social economy. When a hospital dismisses workers who deal with patient inquiries, replacing them with automated systems, it seems to introduce an arrangement that is more efficient. We cannot rely on some "logic of the Digital Revolution" to simply reverse this trend. We must view ourselves as empowered to reject the general direction in which our current efficiency fetish is taking us. If we continue to devalue people who make their livings dealing directly with other people, then we risk ending up with some inhuman digital dystopia. But a genuinely social-digital economy is both possible—it is compatible with the digital package—and worth fighting for.

It is important that we resist viewing any attractive depiction of the Digital Age as a prediction. Predictions, both dystopian and utopian, can be demotivating. Consider this example from history. Between December 22, 1941, and January 14, 1942, Winston Churchill and Franklin D. Roosevelt met in Washington, DC, to decide the general goals and terms of cooperation required to beat the Axis powers. This meeting was a great success. But there was something about their attitudes toward the future that helped it to be such. Churchill and Roosevelt avoided two types of demotivating confidence about the future. First, they avoided a demotivating optimism. When some of today's historians compare the combined productive capacities of the Allied and Axis powers they tend to arrive at the conclusion that so long as the United States, Soviet Union, and British Commonwealth stayed the course then the eventual defeat of Germany, Japan, and Italy was inevitable.[9] It would have been tragically demotivating for Churchill and Roosevelt to bring this belief into their Washington meeting. If they view victory as inevitable then why not raid the President's cellars for bottles of Pol Roger, Churchill's favorite champagne. They also needed to avoid a demotivating pessimism about the future. The meeting would not have gone so well had Churchill announced "We in Britain have sought to contend against Hitler's armored legions. There is no force on Earth that can defeat them. And now they are joined by the invincible Japanese. Where's your Pol Roger?" They viewed the choices that they were making as making the future. They believed that bad decisions would lead to bad outcomes.

Churchill and Roosevelt approached the outcome of World War II with an attitude of empowered uncertainty. They did not place too much credence on predictions about its outcome. But they were certainly not indifferent to these predictions. They understood the magnitude of the challenge that they confronted. We should approach the Digital Age with a Churchillian and Rooseveltian empowered uncertainty. There are many kinds of societies that are compatible with the digital package. Which one we end up with depends on the choices we make and the effort we are prepared to put in. There's nothing inevitable about the social consequences of the Digital Revolution.

Technological revolutions are times of great uncertainty. The desire to predict the future is understandable, especially at times of great collective anxiety. But there's another side to this uncertainty. Technological revolutions are also times of great opportunity. Change is forced upon us. But we can influence *how* we change. As we enter the Digital Age we can think of ourselves as akin to immigrants to a new land. The religious dissenters who boarded the *Mayflower* left an England intolerant of their forms of religion. Emigration to America brought an opportunity to found new kinds of societies. New arrivals to a Digital Age can fecklessly slide into the arrangement described by Andreessen in which a select few get to give orders to computers that transmit those orders to the rest of us. We could tamely permit our age's inequalities to infect and be amplified by the fabulous technologies of the Digital Age. Or we could make creative use of the upheavals of a technological revolution to build societies properly tolerant of human needs. My confidence that we will achieve a social-digital economy is of the very general type advanced by Martin Luther King Jr. when he said "The arc of the moral universe is long, but it bends toward justice." King's statement leaves plenty of scope for significant kinks toward injustice in that arc.

We should be alert to a bias in the way that we think about technological solutions and the way that we think about solutions that may emerge when we consider social solutions, solutions that require people to work together to achieve an important shared goal. If we are convinced of the possibility of a technological solution then we tend not to be discouraged by a setback. We expect that prototypes of moon landers may explode, and that prototypes of driverless cars may fail to distinguish the whiteness of a truck from the whiteness of a spring sky. We feel confident in the arc of technological progress, toward successful moon landings and safe driverless cars. The glitches will be ironed out. We tend not to be so optimistic about fixes for failures of human collaboration. When we fail to achieve a shared goal because some of us were lazy or dishonest we tend to attribute these failures to human nature. We sigh and say that people will always cheat the system when they can. They will take advantage of social welfare programs by claiming benefits to which they are not entitled. Those who have faith in the capacity of humans to sacrifice short-term selfish interests to achieve a shared goal should draw inspiration from cases in which such sacrifices were made. We rightly celebrate the heroic sacrifices made in the war against fascism. There is no reason that such sacrifices could not occur when our shared ends are constructive rather than destructive.

We Should Prefer Robust Ideals

When we describe an ideal we are not making a prediction. An ideal presents a possible arrangement as attractive. To be worth fighting for, an ideal must be more than just attractive—it must be realistic. We must see in it an indication of how to go from our current arrangement to the arrangement described by the ideal. This realism is what tempers my optimism about our collective digital future. A tempered optimism directs that we prefer ideals that we can get to from where we are now.

There's a gap between a lovely sentiment and a practical ideal. Suppose Beatles fans were to offer the famous lyric "All you need is love" as an ideal that could guide resolution of the longstanding disputes between North and South Korea or between Israelis and Palestinians. The lyric seems to describe an attractive way for humans to relate to each other. An imagined arrangement in which all Israelis and Palestinians abruptly came to feel love for each other would lead to a swift solution to many problems. But "All you need is love" is not worth much as advice for those seeking solutions to these longstanding disputes. There's no indication of how to replace the current intense mutual suspicion with attitudes of unquestioning love. Playing the Beatles across the Korean demilitarized zone or into the Gaza strip is likely to be interpreted as crude Western propaganda.

Some ideals are fragile. They are set up so that partial compliance yields none or very few of the ideal's promised rewards. Partial compliance may even reverse the moral polarity of the rewards brought by full compliance. In effect, there is no way to get from partial to full compliance. If you could magically arrive at a situation in which both would-be terrorists and potential victims jointly enjoy endorphins generated by group hugs, then you would incontrovertibly have made the world a better place. But what we know about the world and human nature suggests that it is impossible to get there from where we are now. The ideal's fragility makes pursuit of it very difficult. Partial compliance is likely to produce outcomes that are, in moral terms, the opposite of what was sought. The first ISIS massacre of aspiring group huggers will decisively set back the goals of those who propound the Beatles ideal of universal love. What may be an appealing sentiment offers no waypoints toward its achievement.

Other ideals are robust. They can start small and become more influential. The benefits increase as greater numbers endorse and comply with the ideal, but benefits are still noticeable at low levels of support. The social-digital economy can start small. We can take baby steps toward adequately rewarding people for their generation of social goods and meeting of social needs. Unlike the Beatles ideal, partial compliance generates some of the advertised benefits in ways that can become apparent to those yet to be convinced.

The ideal of the social-digital economy offers waypoints on the voyage to full compliance. As we sustain humans in jobs that might otherwise have been awarded to machines we can realize and enjoy the benefits of dealing with humans. We can draw encouragement from these successes. We can be encouraged to try harder to increase the magnitude of the benefits. Others can witness the benefits received by the early adopters.

It what follows I compare the social-digital economy with two competing optimistic takes on the Digital Age. Which of these visions of our digital future is worth fighting for? The first of these alternatives is Jeremy Rifkin's suggestion of a Collaborative Commons. We will combine our creativity with powerful digital technologies. We will become prosumer members of the maker economy. The second alternative is the idea of a Universal Basic Income. Many citizens of the Digital Age should take advantage of the expected efficiencies of digital machines to absolve ourselves of work. I argue that the social-digital economy better satisfies the criteria of a tempered optimism about humanity in the Digital Age.

The Social-Digital Economy versus the Collaborative Commons

Rifkin's ideal of a Collaborative Commons is certainly preferable to the view described by Andreessen in which a select few of us order computers around while most of us accept that we belong to the part of humanity that gets ordered around by computers. Rifkin issues a general invitation to use digital technologies to join the Collaborative Commons. We can accept the invitation to use the new technologies to connect with each other and enjoy the new collaborative varieties of creativity. Chris Anderson, a former editor of *Wired* magazine, has done much to popularize the maker movement.[10] In his vision of the future, a new class of creative types use 3D printing and digital design tools to launch a new, individual-centered, industrial revolution.

Anderson's writing conveys some of the excitement that accompanies these new digitally enhanced forms of creation. But it's not for everyone. One of the great insights of liberal political thought is the plurality of values that characterize the human condition. Successful liberal societies nourish the different visions of the good life centered on poetry, ultra-marathons, and real estate development. From this liberal pluralist perspective, Rifkin's ideal seems too partial, too specific to digital technologies. Some people just aren't particularly into the Internet and its associated digital advances. They shouldn't be required to locate them at the center of what they do. These people will use email and make Google searches but they will view them as many today view electricity. There are fascinating puzzles in how to bring cheap and environmentally friendly electricity to the masses. But many people will rightly view both electricity and the Internet as boring enablers of how they make their livings. There is a difference between ways of life focused on the development of digital technologies and the exploration of their possibilities and ways of life that stand in the same relationship to digital technologies as they do to electricity. A poet may need a word processor and an Internet connection much in the way that she needs electricity to power the light that permits her to continue writing in the evening. Her vision of the good life may, nevertheless, be little different from that of the early 1800s romantic poet John Keats. It differs from that of someone who spends his day coding APIs (application programming interfaces) for Facebook. The steam engine may have made the Industrial Age but the productive efforts of most people were not directly centered on steam power. The farmers and vicars of early nineteenth-century Britain might, if they chose to reflect, acknowledge that steam power was having a transformative effect on their times. But few of them will have spent much time in factories. Many of them wouldn't make regular journeys by train. The way of the maker is great for some. In a just society, we can all benefit from the presence of makers. But many of us just aren't makers—or at least we aren't the kinds of makers that the Digital Revolution has elevated to prominence. Few are capable of the acts of technical creativity of Apple co-founder Steve Wozniak. Try as we might, few have any chance of coming up with the next Facebook.

Another problem is that these maker jobs belong in a realm that is particularly vulnerable to encroachment and absorption by the very advances in the digital technologies that are the focuses of these creative impulses. When digital technologies get good at designing themselves, there will be a reduction of opportunities for Steve Wozniak and his ilk.

The social-digital economy offers a more expansive vision of humans in the Digital Age. The social economy will have its own heroes and heroines. They needn't be any kind of maker. The social-digital economy would contain its share of well-connected prosumers, but many others would find ways to meet human needs that have little or nothing to do with digital technologies. For some people sharing is not a celebrated new digital economy. It's something you do when you bake a cake and bring it into the office. The social-digital economy makes place for people to participate in areas of the social economy that have very little at all to do with the digital package. A counsellor whose job is to help people distressed by the changes to their circumstances wrought by the Digital Revolution is not a prosumer or a maker. But he is a key contributor to the social economy.

The Social-Digital Economy versus a Jobless Future with a Universal Basic Income

I now compare the social-digital economy with another ideal. This is the idea of the Universal Basic Income or UBI. Many advocates of the UBI accept that most humans won't find employment in the Digital Age. They seek to view this as a feature and not a bug of the Digital Revolution. We can resist Andreesen's and Keen's digital dystopias if we ensure than some of the wealth generated by the machines is shared out to the jobless. Writing in the *Guardian*, Jathan Sadowski notes that the UBI has acquired support from the tech community. "UBI becomes a consolation prize for those whose lives are disrupted. Benefits still accrue to the designers and owners of the technologies, but now with less guilt and pushback about the collateral damage."[11]

The debate about the UBI is multifaceted.[12] My interest here is quite specific: Is the UBI an adequate response to the digital package's threat to mind work? This is how Martin Ford presents it in his 2015 book *Rise of the Robots: Technology and the Threat of a Jobless Future*. Ford describes the UBI as "An unconditional basic income is paid to every adult citizen regardless of other income sources."[13] The UBI will feed and clothe people who can't do anything that the machines of the Digital Age won't do better and more cheaply. But it also permits unemployed and unemployable humans to play an even more important role in the economy. Machines make stuff, but they don't buy it. If almost all humans are shunted off to an unemployed penury, then how will they afford to pay for these efficiently produced goods and services?

Ford presents the UBI as a response to an impending problem for the economies of the Digital Age. If the 1 percent who own almost all the machines also have almost all the money, then we should see a sharp dip in aggregate demand. The immense efficiencies of the digital economy won't count for much if there's almost no one with the money to pay for the resulting goods and services. There is greater aggregate demand in an economy in which everyone has some money than in an economy in which almost all the wealth is in the 1 percent or 0.1 percent but everyone else barely survives on food stamps. Economists have long known that poor people spend a greater percentage of their income than do rich people. Attempts to use tax cuts to stimulate the economy are likely to be more successful if not directed at the rich. A universal basic income permits those at the bottom of the economic pyramid to buy the things that the machines will churn out. The UBI will permit future citizens to perform roles overlooked by too narrow a focus on human economic contributions.

This doesn't mean that strolling malls and watching TV will be the totality of the activities performed by jobless recipients of the UBI. Members of the super-rich class who have tired of making money generally don't choose lifestyles of indolence. Some of them commit themselves to worthy causes. They work to raise money for people orphaned by AIDS or to combat discrimination. We should distinguish work from the institution of work.[14] The institution of work is a feature of today's technologically advanced liberal democracies. We work and expect payment in return. But we all do work for which we don't expect payment. If you spend your weekend fixing a fence on your property you don't expect to get paid. But it shares many of the valuable features—the sense of purpose and challenge—of many of the activities encompassed by the institution of work. Digital Age societies with a UBI may have a very attenuated institution of work. But the citizens of these societies will work. Those limited to the basic income will lack the resources available to Bill Gates, but they can share his charitable goals. They might spend the mornings strolling the malls to discharge their obligation in respect of the economy to help maintain aggregate demand. They can spend their afternoons mounting campaigns for worthy causes.

Paid work will continue to be an option for some. The basic income need not take away incentives to work and earn greater sums of money. Many may choose to spend their days strolling malls but others will enjoy the extra spending power a job will give them. The basic income will suffice for a Toyota, but you'll need paid work if you want a Lexus.

The transition from an economy in which the contributions of many different categories of human worker essential to the economy to one in which only a few of these categories of worker remain could be messy. Technological unemployment resulting from the Digital Revolution will affect some industries before others. Driverless trucks may be safely traversing intercity routes while drivers are still required for taxis that navigate the chaos of inner cities. It could be bad news for the economy if drivers of taxis take the escape route from employment that the state seeks to make available for long-distance truck drivers. If Ford is right, then both categories of job are headed for extinction. But there could be an awkward series of transitions from a situation in which the economy requires the contributions of both kinds of worker, through a situation in which many taxi drivers must be motivated to keep turning up to work while long distance truck drivers should be reallocated to the malls to spend their basic incomes on appropriately priced consumer durables, to an ultimate destination in which neither kind of driver is required. The unemployed existence available to truck drivers cannot be so blissful as to make taxi drivers prematurely extinct.

The UBI as an Inadequate Response to Inequality in the Digital Age

My concern is about the capacity of a UBI to generate massive inequality. We should distinguish the immediate consequences of introducing a UBI from its long-term implications for the Digital Age. Suppose a wealthy, technologically advanced society of the early twenty-first century resolves to offer a UBI to all its citizens. We would witness a quite significant one-off reduction of inequality. But this one-off reduction is likely to be followed by a variety of inequality that we should expect to last longer and to be especially recalcitrant. A UBI is likely to generate a two-tiered society. On top, will be the few who possess marketable skills in an age of powerful digital machines and those who have some significant ownership stake in machines. The members of this group will receive their basic incomes plus wages or rents on the machines that they own. The rest of us will get by on our basic incomes. We should expect significant reductions in social mobility. The kinds of attributes that today suffice to bring wealth to someone born into poverty are significantly less likely to have this effect in an age in which much mind work is done by machines.

First, a general observation about inequality. We oversimplify the current problem of inequality when we think of it as the problem of the gap between the haves and the have-nots. It is better characterized as the gap between the have-lots and the have-littles. Many of today's poor have possessions that would be envied by the rich of yesteryear. They have color televisions. They enjoy hot showers. They treat their sore throats with antibiotics. Some of the poor are hungry. But in wealthy liberal democracies complaints are more likely to be about the quality of food available to the poor. The poor don't starve, but they have bad diets. Parents rushing between a variety of badly paid part-time jobs have little option but to feed their children junk food.

Problems arise with respect to the large gaps between the wealth of those at the top and those at the bottom. It's one thing to be given sufficient money to feed and clothe a family, but we have needs beyond these basic necessities. These needs are significantly influenced by our positions in society relative to others. Arrangements that we judge to be egalitarian need not aspire to reduce the gap between the best off and the worst off to zero. There are better off and worse off people in societies that we judge to be egalitarian. But egalitarian arrangements seek to minimize this gap in ways that are compatible with the basic rights of citizens and the need to generate prosperity.

The UBI is well designed to prevent those at the bottom from starving and keep them supplied with things from the dollar store equivalents of Digital Age malls. But a UBI motivated by the need to maintain demand is likely to widen the gap between the have-lots and have-littles. Social mobility is a valuable feature of a society in which benefits are distributed unequally. We celebrate stories like that of Andrew Carnegie, the son of a sporadically employed immigrant weaver whose talent and determination led to extreme wealth and fame. It is hard to see how a society in the Digital Age that pays a UBI could preserve this path from the socioeconomic bottom to the top. Life should be significantly better for Digital Age recipients of a basic income than it was for William Carnegie—Andrew's father. But the great proficiency at mind work of Digital Age machines means that very few will get to climb the ladder scaled by Andrew. Very few people born into the class of people who get by on the basic income alone will discover that they have marketable skills in an age of superlative machine learners. Some Digital Age Larry Davids and Oprah Winfreys will discover that they have talents that are truly remarkable and beyond the reach of the AIs. They may thereby ascend to society's upper rungs. But the path of working hard to acquire capital will be less available to those who hope to rely on hard work. Most of the unemployed class will get by on basic incomes that feed, clothe, and entertain them quite well, but won't suffice to acquire a significant ownership stake in the digital economy.

The advance of digital technologies threatens to negate the salutary function of education in a liberal society of promoting social mobility. As machine learners colonize the domains of activity in which the children of the poor might have displayed their talents, the paths of escape should predictably narrow. We will accept that position is inherited. What matters here is not the inheritance of genes with putative links to worthy traits. People will be less interested in claiming to work hard and to have exceptional business acumen once it's clear that these attributes make less of a difference than the ownership stakes in the economy you inherit. We could return to a rentier society in which what you inherit matters much more than anything you do. In Marxian analysis, the rentier class is defined by significant holdings in property that generates profit. The members of this class have no need to make social contributions. They can maintain an indifference toward the activities of the ventures in which they have holdings. They care only about the monthly check generated by these ventures.

Participation in the labor force offers a further means of improving conditions that will become less available in a workless Digital Age. Workers who feel ill treated can threaten to withdraw their labor. If wealthy New Yorkers want to continue drinking soy mocchaccinos, they must adequately remunerate those who make them. They want the people who fill these roles to be sufficiently well paid to motivate them to keep on turning up for work. If those who make soy mocchaccinos judge pay and conditions to be inadequate they can threaten the withdrawal of their labor. The future jobless of a society in which those jobs are done by uncomplaining digital technologies won't have this historically important means of complaint. The have-lots are unlikely to take time out from sipping their machine-made mocchaccinos to read editorials in *The Big Issue* complaining about the conditions of those who get by on the basic income alone.

What about the important economic role described by Ford—that of consumption? In Ford's conception of the Digital Age, the economies of the Digital Age face ruin if the jobless stop buying. The jobless of Ford's Digital Age can exercise considerable influence by their choices of which basic items to purchase. Manufacturers of breakfast cereals should invest considerable sums of money to make their brands attractive to the jobless. But it's hard to see how those who get by on the basic income alone could use their choices over what to consume to register a more fundamental complaint about the status quo. A threat to withdraw the economic service of consumption is effectively a threat to starve. It is, for that reason, not particularly credible. We remember people who took the option of self-immolation to register protests about war in Vietnam. But that avenue of protest takes a degree of commitment shared by few. In the Digital Age, it would seem to be a necessarily self-limiting way to signal disapproval of your permanent relegation to the consumer class.

Those at society's bottom tiers may have legitimate complaints. Consider, for example, the proposal of the Belgian philosopher Phillipe van Parijs. He argues that the basic income should be set at maximum level sustainable by a society. Van Parijs justifies this level in terms of a commitment to produce the most "real freedom" in a society.[15] According to van Parijs, your level of real freedom is set by the actions that you can perform. Put simply, if *A* can do more things than *B* then *A* has more real freedom than *B*. Real freedom can be limited by laws prohibiting certain acts but also by a lack of resources to perform those actions. You may be happy to not live in a society in which it is illegal for you to drive a Lexus, but poverty can make you just as unfree to purchase one as a legal ban. Van Parijs argues that we should seek to maximize the real freedom of members of society that possess the lowest levels of it.

This is an excellent philosophical proposal. But there are practical obstacles blocking its realization in a society in which machines do much of the work. Currently, the worst off can supplement their philosophical arguments for better treatment with the threat of withdrawing their labor. Consider the comparison with the abolition of slavery. The threat of violent revolt by enslaved people may not be required by any premise of a sound philosophical argument for the immorality of slavery. But the threat of revolt gives forceful expression to those arguments. In a Digital Age in which the worst off don't work they must hope that sufficiently many of their socioeconomic betters take the time to encounter and be convinced by van Parijs's arguments. They may be as likely to be convinced as slave holders whose human chattels write powerful editorials rejecting the condition of slavery, but promptly comply with every command issued by their masters. They may need the arc of the moral universe to be very long before polite philosophical argument has the desired effect on slave owners.

An Expanded Basic Income?

Perhaps we make a mistake if we conceive of basic incomes as limited to a fairly basic collection of goods. If an interest in maintaining consumption can keep the dollar store well stocked, then why shouldn't the same reasoning apply also to the thousand-dollar store?

There are currently trials of the UBI that involve comparatively small payments. In his recent defense of the UBI the Canadian philosopher Mark Walker suggests a specific sum as sufficient to meet the basic needs of Americans—US $10,000 per annum.[16] He presents this sum as amply covered by the capacity of the US economy to generate wealth once appropriate cuts are made to comparatively unimportant expenditures such as the military. Walker grants that US $10,000 per annum suffices only for a frugal existence. But if we are interested in the long view of the Digital Age we must take account not only of the form the basic income would take when first introduced, but what it may become. Perhaps the same economic justification for a basic income that covers a frugal existence might be extended and applied to more expensive goods? If free money is a useful spur to the production of cheap Toyotas, then why shouldn't the same reasoning be applied to luxury Mercedes Benzes? This process could be abetted by technological advancements. The techno-optimist writer Byron Reece enthuses about advances in manufacturing that will permit a Mercedes Benz to be built for $50.[17]

I suspect that we are unlikely to see the expansion of the basic income beyond a quite circumscribed collection of basic goods. This is because the goods purchased by the rich are valued by them, in part, because of their inaccessibility to those with more modest means. No one wants a $50 Mercedes Benz. The Mercedes is a luxury product and the demand for luxury goods differs from demand for their nonluxury counterparts. An objective assessment of a cheap car such as the Toyota Corolla and a luxury car such as a Mercedes reveals little difference between them. Both can safely, comfortably, and efficiently take passengers from point *A* to point *B*. Given the functional similarities of the Toyota and the Mercedes it becomes apparent that much of the additional value of the latter is as an indicator of social status. It serves this purpose by not being available to all. A $50 Mercedes sends none of the signals of enhanced status sent by a $100,000 Mercedes. The Mercedes has some features lacked by a cheaper car, but much of its value to its owner lies in its broadcast of success. If you are considering a $50 Mercedes, then why not pay $25 for the basically as good Toyota? The possession of neither car supports a claim to membership of society's elite.

It wouldn't matter so much if the class of goods available only to the rich were restricted to those that confer enhanced social standing. The poor get Toyotas; the rich get Mercedes. The poor get iPhones; the rich get gold-plated iPhones. The members of both groups get to visit the same locations and enjoy the same smartphone features. But the category of goods available only to the rich is likely to extend beyond those that merely confer enhanced social standing. It will include the new medical and educational technologies, technologies that both enhance social standing and make more concrete differences to how well life goes. Those who either own the machines that do mind work, or do work that these machines cannot do, will want to believe that they get more than gold-plated versions of the technologies available to the jobless. They will insist that, beyond signaling success, these technologies make significant differences to how they live their lives. They will want rejuvenation technologies for themselves and enhancement technologies for their children.

The digital future likely to result from Ford's suggestion will be one in which there is a sharp distinction between those who receive nothing beyond the basic income and those who receive the basic income plus payment for the exercise of some skill that retains value in a largely automated future and those who have a significant ownership stake in the machines.

I predict that the have-lots will be especially interested in ways to distinguish themselves from the have-littles. Those who receive nothing beyond the basic income will be expected to stroll the malls. But their purchasing activities are likely to be restricted to items on the lower shelves. The have-lots will not be at all interested in these items. If they do turn up to the same malls they are likely to insist on being taken to the private viewing rooms where the most exclusive items are available. These exclusive items should make the lives of the have-lots very different from the lives of the have-littles.

Concluding Comments

In this chapter I discuss what is involved in offering an ideal for the Digital Age. Ideals are not predictions. Good ideals for the Digital Age are attractive and achievable ways for humans to inhabit it. We must approach their realization with an empowered uncertainty. Whether we achieve them depends on what we do. Treating them as inevitable outcomes of progress in digital technologies will reduce the likelihood that we will enjoy the benefits of the social-digital economy. I compare the social-digital economy with two other attractive Digital Age ideals. Rifkin's ideal of a collaborative commons in which people use digital technologies to share and create value seems too specific to the operation of digital technologies to be a general ideal for the Digital Age. Some people will enjoy working with others to create beneficial digital novelties. Others will lack those skills and interests. They will seek work in an expanding social economy. The ideal of a universal basic income promises a one-off reduction in inequality. But it threatens to bring a replacement variety of inequality that is especially recalcitrant. We should expect low levels of social mobility in a society whose machines do much of its mind work. Those without work in a Digital Age with a UBI will be denied the great benefits brought by a vocation.

8 Machine Breaking for the Digital Age

Calls for a more moderate engagement with digital technologies tend to attract accusations of "Luddite!" This insult is inspired by Ned Ludd, a textile apprentice whose smashing of a machine in the late 1700s made him a symbol for workers fearful for their prospects in a time of rapid technological change. The lesson from history is supposed to be straightforward. You can moan all you like, but you can't stop progress. The prefix "neo-" repurposes the insult for the Digital Revolution. The Luddites didn't stop the Industrial Revolution; Neo-Luddites won't stop the Digital Revolution.

Some historians offer a more sympathetic way to think about the Luddites.[1] Rather than viewing them as ill-informed fantasists who failed to stop the Industrial Revolution, we can think of them as demanding a fairer share of the bounty of technological progress than that initially offered by those who built the factories and installed the power looms. Luddite machine breaking can be placed in a context that includes other forms of worker protest that led, in the end, to worker protections and entitlements. My aim here is certainly not to stop the Digital Revolution, but rather to influence how it unfolds. Decisions that we take now set powerful precedents for the Digital Age.

I'm not suggesting that anyone show up to one of Amazon's vast data centers with a pickaxe. What's called for is a more genteel form of digital machine breaking that makes use of market incentives. We've seen how Facebook benefits from the "billion times anything" principle. But even a small general decrease in enthusiasm about a company that is perceived as profiting unfairly can affect profits in ways that Facebook notices. We can communicate our displeasure to Facebook, Google, and the rest in ways that matter to them. We can take steps toward realizing the ideal of a genuinely social economy. We can engage in forms of protest that may lead to a more human Digital Age with a smaller—but still considerable—share of its economic returns going to those who own the machines.

See through the Digital Halo Effect!

We must do something about the halo effect enjoyed by some of the central actors in the corporate tech world. We treat the big tech companies differently from other corporate heavyweights. We have long experience of the kinds of moral shortcuts that oil and gas multinationals take in their accumulation of wealth. We tend to treat some of the leaders of the corporate tech world as accidental billionaires. The proclamations of tech pioneers that their purpose is to "make the world a better place" were well satirized in the HBO comedy *Silicon Valley.* In February 2017, Mark Zuckerberg published a 5,700-word letter addressed to the Facebook community entitled "Building a Global Community." It was promptly dubbed the "Zuckerberg manifesto."[2] In it, Zuckerberg tells us "Facebook stands for bringing us closer together and building a global community." He says of Facebook's moral mission "the most important thing we at Facebook can do is develop the social infrastructure to give people the power to build a global community that works for all of us." Airbnb has rapidly acquired a market valuation greater than traditional hotel industry giants Hyatt, Marriott, and Hilton.[3] This focus has not prevented an Airbnb insider expressing an aspiration to someday win the Nobel Peace Prize for "helping with cross-cultural understanding."[4] We credit some digital economy billionaires who are less voluble about being moral with a selfless desire to create beautiful things. Steve Jobs faced criticism for not giving much to charity.[5] The digital halo effect gave him a pass. The Steve Jobs legend presents him as a Digital Age Michelangelo, the iPhone as his David. Jobs was all about the exquisite design and not about the money. Beautiful objects were his way of making "a dent in the Universe."[6]

What should we make of these proclamations of moral or aesthetic ambitions? Moral proclamations come with implied costs. They suggest a willingness to accept the costs of promoting their ends. It's easy to wish for a world in which no one is hungry. But a world without hunger is a moral goal for you only insofar as you are prepared to do something to achieve it. Another possibility is that the titans of tech are operating in ways that

Milton Friedman would applaud. They are using moral language to cloak their real mission to maximize shareholder value.

A good test of the commitment of the titans of tech to the goal of making the world a better place is seeing how they act when the goal of making the world a better place conflicts with the goal of making money. The Zuckerberg manifesto acknowledges the challenge of fake news in the lead-up to the 2016 US election. Fake news colonized many Facebook newsfeeds. But the money Facebook makes from targeted advertising is unaffected by whether members are reading accurate reports on the gravity of the international refugee crisis or fake news reports of Pope Francis's endorsement of Donald Trump's presidential bid. If the desire to make money is paramount, then expect Facebook to avoid any measure that would reduce its user engagement. If this causes complaint and threats of government regulation, then the Friedman strategy of offering moral pronouncements to cloak actual indifference is available. One measure of relative importance Facebook places on its touted moral mission and its accumulation of profit are the measures it takes to avoid paying tax. An *Observer* editorial on the Zuckerberg manifesto identifies the gap between its moral proclamations and its actions: "Like other corporate giants, Facebook has done all in its power to minimize its tax bill, paying a fraction of what it owes the societies from which it draws its profits, undermining the very social infrastructure Zuckerberg claims to want to build."[7] The motive to make money seems to explain much of what the titans of tech do. Amazon's Jeff Bezos didn't start out as a book lover dreaming of new ways to spread the joy of reading who made the serendipitous discovery that a platform for selling books could be used to sell other things.[8] He approached books much as Exxon Mobil approaches oil—as a commodity. If Bezos had expected more money out of fracking oil than selling discount copies of Leo Tolstoy's *Anna Karenina,* then he would have done that instead.

This is no critique of capitalism. Capitalist societies need businesses whose principal purpose is to make money for their shareholders. But it is important that we recognize that when they say they want to make the world a better place, they are diverting our attention from their real focus on profit. There are many organizations—for example, the United Nations and the Intergovernmental Panel on Climate Change—that can be credited with an authentic desire to make the world a better place. You might have doubts about the effectiveness of the UN or the IPCC at making the world a better place, but they seem a better bet if we want organizations that promote our moral interests than are tech companies that seek moral interests only to the extent that they do not conflict with the maximization of profit. When the titans of tech do as Bill Gates did and install themselves as chairs of charitable foundations we might take them at their word. The Bill and Melinda Gates Foundation has a more serious interest in alleviating global poverty than Microsoft Corporation ever could.

It's important that we not delay in our reassessment of the motives of the leaders of the digital economy.[9] We should expect to see a vast increase in the profits of companies that hold data as they expand the reach of machine learning. Google and Facebook are like Standard Oil before the commercialization of the internal combustion engine. If you think that Amazon, Apple, Google, and Facebook are wealthy companies now, then you should expect to revise up that assessment when they begin to extend their reach in to areas that Robert Gordon identifies as strongly affected by the Second Industrial Revolution: food, clothing, housing, transportation, health, medicine, and working conditions but, thus far, relatively untouched by the digital revolution. Now is the time for the kinds of democratically inspired intervention that in the 1910s broke up Standard Oil. We should act before Google, Facebook, and Co. consign democratic institutions to the same trash can of history that contains Netscape and AltaVista.

We shouldn't be tormented by fears that expecting Apple and Google to pay their taxes might kill the goose that laid the golden egg. The Russian oligarchs made the most of fortuitous locations in relation to the natural resources of the Soviet Union as it disintegrated and its component republics made halting moves toward free market capitalism. Page, Brin, Bezos, Zuckerberg, and Co. found themselves in a similarly fortuitous location in relation to the wealth generated by the digital package. Amanda Schaffer writes about the influence of "great man" myth on our understanding of the Digital Revolution. She says "Rather than placing tech leaders on a pedestal, we should put their successes in context, acknowledging the role of government not only as a supporter of basic science but as a partner for new ventures."[10] Our current titans of tech are very intelligent and gifted individuals. But they are replaceable. They may sulk if we demand that they pay proper taxes, but they don't have the option of taking their toys and going home.

We should also recognize that the frequency of the use of the term "tech billionaire" to describe those who we see as driving the digital economy is, to some extent, normalizing a very unequal distribution of the wealth generated by the digital package. There's nothing about the digital package that requires that the founders of and early investors in technology companies become billionaires many times over and that many of the rest of us find our skills increasingly devalued. We should expect the Digital Age to contain big technology companies headed by some very rich men and women. But we could seek to revert to a quaint and retro conception of great wealth in which the founders of Facebook and Google are mere multimillionaires and not multibillionaires. Tech multimillionaires would have millions and millions of dollars, enough money to own mansions in nice parts of town and cottages in the country, and to buy first class air tickets to any grand slam tennis tournament, but they wouldn't have enough money to compete with NASA in the space exploration business. This may seem like a fanciful vision of our collective future and perhaps it is. But the thing that makes it fanciful is a collective unwillingness to mess with our current political and fiscal arrangements, not some irresistible logic of the digital package.

Don't Fall for Tech TINA!

TINA is an acronym for "There Is No Alternative." It was a slogan promulgated in the 1980s by Margaret Thatcher, the British Conservative Prime Minister that suggests a futility in resisting the dictates of the market. We may want the operations of railways to conform to our social priorities, continuing to service remote communities, but we are powerless to resist the directives of the market in determining whether and how such communities are served.

TINA illustrates the powerful effect of assumptions about how the world works on our collective consideration of alternatives. Once you decide to consider policies in market terms, some possibilities become difficult to see. Recognizing them requires you to question psychologically entrenched assumptions. There is a technological version of TINA that has the same constricting implications on our consideration of alternatives as Thatcher's economic version. In tech TINA, some choices become irresistible outcomes of technological progress. We can hope to delay the inevitable. But such delays place us at a disadvantage with respect to the benefits of technological progress.

Tech TINA directs that we evaluate the work done by humans purely in terms of efficiency. Humans must produce more widgets more cheaply to be viewed as more efficient than machines. I have suggested that this omits some of the most important contributions of human workers. When we interact with human workers we interact with beings who have minds like ours. We don't have to see this interaction as more valuable than the outcomes of the interaction—the meals delivered and the medicines dispensed. The number of meals delivered and the medicines dispensed may be the most important things. But we can still see humans as contributing something that is worth preserving. They contribute their distinctive mind labor.

We should beware of presentations of wealth inequality as inevitable consequences of the digital package. The industrial package offered choices including the path of Gilded Age America, the path of the 1930s Soviet Union, and the path of 1970s social democratic Sweden. The decisions we make now set powerful precedents. We might imagine Ukrainian victims of Stalin's forced collectivization and industrialization accepting a Second Industrial Revolution version of tech TINA—their suffering, regrettable though it may be, was part of the price of industrialization. But it really wasn't.

One way to resist tech TINA is to insist on dealing with comparatively inefficient humans whenever possible. In these times of economic austerity businesses are shedding staff. I have argued that this overlooks the value of human social contributions. There are ways to send market signals that you like interacting with humans in contexts in which the social goods that they generate are properly valued. If you are dealing with some corporation's helpline to get help with one of its flawed products, then repeatedly pressing "0" on your touch-tone telephone might get you through to a human. When you seek out a human employee you aren't necessarily seeking to date them. Humans are obligatorily gregarious animals who enjoy even the most fleeting pleasures of interacting with other human animals. Choose supermarket checkouts staffed by humans. The role of supermarket checkout worker in 2018 is no one's dream job. But today's version of that job could give way to a socially-enhanced version. Harrison Ford is well paid for injecting his humanity into his movies. We recognize the value of that humanity when we pay to see his movies. Socially enhanced supermarket checkout workers should have a similar expectation to be paid for exercises of their social skills. Customers should be prepared to pay just a bit more for goods sold to them by socially enhanced checkout workers.

There is no reason that socially enhanced sales assistants should not thrive in the social economies of the Digital Age. A socially enhanced sales assistant helps shoppers to choose products that are right for them based on their own and reported experiences of products. We have seen this separation in the restaurant business. When people visit a fast food shop they are typically motivated by efficiency. They want cheap cooked food. But when they visit a sit-down restaurant they are aware that they are paying for the social aspects of the job.

The workers that automating supermarkets are letting go may be told that the Digital Revolution will bring jobs more befitting of their moral status. That may be so—it's difficult to predict the jobs that will emerge from the Digital Revolution. But in the meantime, we should accept that the fact that supermarket workers turn up to work indicates a preference for that job over no job. Supermarket checkout workers are not different from corporate lawyers and academic philosophers in being able to imagine possible jobs that are preferable to those they currently have. When you choose a checkout with a human worker you aren't degrading him or her.

If You Can Cheat an Algorithm, Then Why Not?

We've seen the challenge from machine learning to human mind work. If your job is to identify malignant lesions in ultrasound images, then you are doomed. Machine learners will predictably do what you do to a much higher standard and more cheaply. Humans cannot win these competitions against machine learners. Other competitions pit people not against the machines themselves, but instead against the humans who own the machines and profit from their operation. You may not be able to outthink the machine. But perhaps you can outthink the machine's owner.

Young people seem to be those who are getting the raw end of the deal as our society transitions to the Digital Age. They are told not to expect the easy employment and strong worker entitlements enjoyed by their parents. Young people can use their superior understanding of digital technologies to respond to this intergenerational injustice. They are not seeking to outthink the machines but instead to outthink their human elders who exercise a disproportionate control over digital technologies and acquire a disproportionate share of the wealth.

When parents seek to deny their children access to the liquor cabinet they can use a key to lock it. The children are unlikely to possess an understanding of the workings of the lock that exceeds that of their parents. But, as Mary Aiken notes, young people have a better understanding of the Internet than that of their parental would-be regulators.[11] The same differential in technical understanding limits the older people who make laws that protect their claims on the wealth produced by the digital package. Younger people have an advantage here. They are digital natives who can make use of the digital package in ways that older people did not anticipate. They should feel empowered to exploit gaps in the regulations set by their elders.

Uncertainty about what is possible for the digital package by those who seek to make the rules means that these young people will often spot loopholes—behavior that those who made the rules might have banned were they to have been sufficiently imaginative to think of them. One such example is the peer-to-peer file sharing enabled by Napster. Suppose you want to distribute copyrighted material. You can't store it on a central server. That server becomes the target of action by the lawyers charged with protecting the interests of its legally recognized owners. Peer-to-peer file sharing obviates the need for a central server to host pirated material. Pirated material lives on the many computers linked by file-sharing software.

It's appropriate to feel bad about cheating other people. If a human sales assistant fails to charge you for an item, then it is appropriate to point out the error. You should think about the feeling of betrayal felt by a human being when she provides a service that you fail to pay for. When a poorly coded corporate webpage permits you to benefit in unanticipated ways, then why not make them pay the price for dispensing with human workers? The norms of human decency do not apply to our treatment of machines.

Of course, some distance behind the algorithms are human beings who are made worse off in the same kind of way that the human sales assistant who forgets to charge you. Milton Friedman argued that the pursuit of profit by businesses should conform to the "basic rules of the society, both those embodied in law and those embodied in ethical custom."[12] The coders of algorithms may typically seek to comply with the letter of the law. And so should aggrieved, technologically savvy young people. We should understand that when a new technological package is introduced there is considerable uncertainty about how laws and other "basic rules of society" apply to its novel ways of generating wealth. It often seems to be the case that the law is forced to play catch-up as the technologically aware discover new ways to access content that companies claim a proprietary interest in.

Work for Free for Oxfam, but Make Facebook Pay!

In chapter 3 I described Jaron Lanier's plan to introduce micropayments. I suggested that our cluelessness about the digital package made these a difficult sell. Today's technology companies benefit from the kinds of action that bring zero or negligible benefits prior to the Digital Revolution. When Fitbit asks you to choose between two innocuous options "I found this article helpful" and "I did not find this article helpful," you should realize that it's doing something different from what a human phone sales representative does when he asks you "How's the weather with you?" That question is motivated by a fleeting feeling of curiosity about the weather in Wellington, New Zealand. Fitbit's alternatives satisfy no feeling of curiosity. They are about gathering data. It's rude to just ignore the inquiry of the human sales representative. It's appropriate to ignore Fitbit's inquiry. Fitbit expects to use your answers to make a better product for you. But you will be expected to pay what Fitbit considers to be the market value of those improvements. Tech giants have profited richly from our tendency to reflexively proffer requested information. Answer if you want to. But be similarly disposed to return petrol that you predict that you won't need to the vending oil company free of charge. It's hard for us to recognize the economic value of the contributions we make to tech companies when we grant them access to our digital exhaust. Lanier describes a redesign of the Net that would be required to implement a universal micropayments system. But we must change first. We must rein in our feelings of gratitude for the freedom to work the digital fields of our digital lords. When considered individually, micropayments are trivial. But they accumulate. We could subscribe to our own version of Facebook's "billion times anything" principle. A thousand times anything may be enough to pay some of your rent. We will be incentivized to improve the quality of our online contributions if we see them as part of the way we make our livings. We can try to fast-forward our psychological and emotional adjustment to the digital package.

Paying closer attention to the digital package could enable the redirection of inner conspiracy theorists away from moon landing conspiracies toward the measures that technology companies use to separate us from our data. There has been much written about the user-unfriendly format of the terms and conditions for technology companies' services. Why do technology companies ask us to assert and reassert that we have read the terms and conditions that apply to our use of their services when almost all of us predictably haven't? The wall of words that greets those interested in any of Apple's services isn't designed to be read and internalized by its customers. This obfuscating legalese comes from a company that has an established record of providing information in a user-friendly way. When doctors elicit our consent to medical procedures they seek to make us understand the implications of our consent. They seek to acquaint us with the odds of success and the implications for us of the procedure's failure. They certainly don't present pages of small print and ask us to sign on the dotted line or click the "I accept" button. Imagine that Apple applied the same genius that helps iTunes customers find music that is just right for them to helping them to grasp all the implications of its terms and conditions agreements.

Don't Fight the Last War!

There's a famous line about generals always fighting the last war. France's Maginot Line, laboriously constructed between 1929 to 1938 might have helped against a German army using the strategies and tactics of 1914. It was an inadequate response to the highly mobile German army of 1940.

We shouldn't fall into the trap of building a Maginot Line to protect against the injustices of the Digital Revolution. It is a mistake to fight the injustices of the Digital Revolution with measures designed to respond to the injustices of the Industrial Age. The labor union is well adapted to the factories of the Industrial Age that brought workers to a single geographical location. Workers could organize on the factory floor and form picket lines outside of the factory gates. Labor unions are not particularly effective at responding to unfairness specific to the digital package.

Consider the example of Uber. We can be thankful for some of the things Uber has done. It has brought cheaper rides and flexible employment. We should nevertheless acknowledge the feelings of acute unfairness it has caused. Uber seems to divide up the wealth it generates in ways that don't seem fair. It is a challenge to established ideas about distributive fairness.

Uber's 2018 valuation of US $72 billion is good news for its investors.[13] The news is not quite so good for Uber's drivers—or, as Uber prefers—its "partners." Investing in Uber may be a way to achieve great wealth, but partnering with it isn't. Uber retains more than 20 percent of the money paid by passengers, leaving drivers to cover their own business-related expenses.

I think we can learn from the Digital Revolution that certain ways of challenging Uber's share of the money paid by passengers and their reluctance to offer traditional worker protections to drivers are more likely to succeed than others. Those who think that Uber doesn't pay its partners enough will do better if they offer suggestions in synch with the digital package than if they insist on the restoration of ideas about distributive fairness tied to the industrial package of technologies. What follows is one suggestion that seeks to turn the power of digital networks against Uber.

In chapter 7, I distinguished ideals from predictions. We may predict that an ideal is especially difficult to realize but consider it worth promoting. I think that the odds are heavily stacked against meaningful collective action on climate change. I nevertheless consider the ideal of a low carbon economy to be sufficiently important to be worth fighting for regardless of my pessimism about its realization. The brief defense of a worker platform that I set out below falls well short of counting as a plan. I offer a couple of tips about how to implement a plan, but nothing that I say should lead to a complacent optimism about the prospects for the disempowered partners of powerful digital platform businesses. Some people may find the ideal I present attractive. If so, they will have their work cut out for them.

One thing that Uber has taught us is that if a well-engineered platform business enters your industry and you confront it as an individual, you won't do well. The platform will suck up almost all the additional wealth it helps to generate. Robert Reich says of the economy we are "barrelling toward": "The euphemism is the 'share' economy. A more accurate term would be the 'share-the-scraps' economy."[14] A problem is that unions—the historically most important instruments of worker protection—find themselves wrong-footed by Uber's exploitation of the digital package. Unions are a legacy of the Industrial Revolution purpose-built for the factory floor. Platform businesses call for new ways to organize workers. Workers are being directed away from expectations of jobs for life. They are joining a gig economy characterized by many temporary positions in which they are expected to act as independent contractors. Workers in the gig economy might drive for Uber in the morning, make deliveries for Taskrabbit in the afternoon, and spend the evening fitting out a spare room to be rented on Airbnb. The gig economy calls for *worker platforms* that mirror key features of the platform businesses that they are built to confront. Uber came into existence as a website and an app. And this is how a ride-sharing worker platform can originate. Uber's profits grow as its user network grows. Worker platforms gain bargaining power as their worker networks expand.

Platform businesses feature low barriers to entry. It's easy and free to sign up to Uber. That is a key feature of its success. Uber does not charge for access to its very valuable network in the way that the *New York Times*'s controversial paywall aspires to make users pay for access to its very valuable journalism. Uber wants its app to find its way onto your smartphone and wouldn't dare charge even a modest fee for it. A ride-sharing worker platform should copy this feature. It should not charge membership dues. Like many platform businesses, it will originate as a website that invites membership very broadly. If the platform business it confronts operates internationally, then the membership of the worker platform must be international too. It shouldn't begin in a particular city and then look to expand.[15] There are Uber drivers in Rabat, Reno, Riyadh, and Rome. The Internet is there too. A worker platform that responds to Uber must be equally available to drivers in all those places. There are many differences between drivers in Rabat, Reno, Riyadh, and Rome, but if they are partners of Uber there is likely to a shared concern about how much of the money paid by passengers trickles down to them.

Platform businesses keep their overheads low. So should worker platforms. They might launch themselves on the crowdfunding site Kickstarter. There they would seek small sums of money sufficient to sustain a focused range of activities. Funding might come from angel investors. Here I mean genuinely "angel" investors, wealthy individuals worried about the effects of economic and technological dislocation on the bottom 90 percent, rather than those with narrow commercial interests in start-up businesses.

Wealthy individuals with a genuine interest in the welfare of those who bear a disproportionate burden of the costs of digital innovation may be rare, but they do exist. Cynics may choose to view the Bill and Melinda Gates Foundation as a clever way to boost the value of Microsoft stock. Another possibility is that Bill Gates has some genuine interest in the well-being of the worst off. He is not engaging in the Friedman-esque strategy of using moral language to mask plans to boost shareholder value. The example of Gates demonstrates that, rare though they may be, beneficent billionaires can use their wealth to do a great deal good.

A worker platform represents the interests of workers most effectively by renouncing any interest in a piece of the action. There are many platform businesses seeking to monetize aspects of the way users interact with each other. The lack of any commercial interest should make the platform more appealing to workers. We expect politicians to be free of conflicting commercial interests. Workers should expect the same of a platform set up to represent them.

Low-overhead worker platforms must limit what they do. They cannot hope to do many of the things done by traditional unions. A ride-sharing worker platform will lack legal teams that can credibly threaten to take Uber to court. It cannot deal with individual complaints about ill treatment. Its central purpose is to get better deals for workers out of platform businesses.

Why should Uber listen to a ride-sharing worker platform? Historically this has been a big problem for unions. It's difficult to argue for workers whose employer refuses to recognize you. Here again there are lessons from platform businesses. The value of a platform business lies principally in its network of users. And that is where the power of a worker platform lies. Uber won't listen to a ride-sharing worker platform with ten members, but it should listen to one with ten thousand. A ride-sharing worker platform stands ready to talk to Uber—and to any incoming ride-sharing platform business that might offer drivers a better deal. It doesn't rely on notions of fairness that are unlikely to get a sympathetic hearing from Uber founder Travis Kalanick, a self-proclaimed devotee of Ayn Rand.[16] The worker platform can use the power of its network to appeal to Kalanick's self-interest. It offers something of great value to an Uber competitor prepared to offer workers a better deal—its entire membership as partners for the new business. Drivers don't need to fight Uber, instead they can subvert it.

Uber is certainly not the only ride-sharing business. It faces competition from Lyft in the US and BlaBlaCar in Europe, for example. Left to their own devices, businesses tend to conform to the ecological principle of competitive exclusion—they find different niches in the ride-sharing economy. Lyft places greater emphasis on community engagement. Uber owns the luxury car market. The interests of workers are best served when Uber and Lyft compete directly. A direct competitor is the ideal recipient of an ill-served worker network.

An emerging worker platform faces some of the same obstacles as a new platform business. It's difficult for an emerging platform business to generate a critical mass of users. The value of a platform increases as the number of users increases. This makes it difficult for platform businesses to get started. A dating site with a million members is much more valuable than one with ten. The problem is how to get from ten to a million. Who wants to join a dating site with ten members? The growth strategy for an emerging worker platform is obvious. It knows exactly where to find its users. It parasitizes the membership of the platform business it seeks to confront. The common themes in driver complaints about Uber form the basis of its pitch.

A worker platform has low barriers to entry. But it requires some degree of commitment from its members. It shares this feature with platform businesses. You aren't much use to Uber if you sign up, download its app, but never click on it. The ride-sharing worker platform needs something from its members too. Those members must show some interest in negotiations on their behalf. They must demonstrate some disposition to act on its advice and to switch partnership to a new ride-sharing platform that offers workers a better deal. They should feel confident that many members of their worker platform will join them in making the switch. Those who negotiate for better deals need to credibly claim to Uber and competing businesses that drivers will act on their advice. Uber can't really complain. The facility with which it sheds unwanted partners suggests a reciprocal freedom to find preferable ride-sharing platform businesses.

Membership of one worker platform does not preclude membership of others or indeed of a traditional union. In the gig economy workers are told that they must be prepared to take on many roles. They would register with all the worker platforms that represent their various gigs.

There's always the fear that an effective worker platform will merely intensify Kalanick's desire to "take the dude out of the front seat" and have

Uber transition to driverless cabs.[17] But Kalanick is already strongly motivated to do that. His interest in margin expansion means that he resents any money paid to drivers. Uber could retain 80 percent of the money paid by passengers and drivers a mere 20 percent and drivers would still, from Kalanick's perspective, with his interest in a driverless future, be getting too much. A worker platform could enable drivers to get more out of the remaining years of employment available to them. It also offers a model for those whose services are essentially social.

We've heard far too many stories about the billionaires made by the Digital Revolution. It's time for a greater focus on how wealth can be spread around rather than allowed to pile up in the bank accounts of a few Internet-appointed masters of the universe. I have suggested one way in which this can occur that pays attention to the mechanics of digital wealth. It would be wrong to minimize the obstacles blocking the realization of worker platforms able to compete on a level footing with Uber and the like. We should not confuse predictions worth wagering on with ideals worth fighting for. The ideal of powerful worker platforms is worth fighting for even if we are pessimistic about its prospects.

Concluding Comments

The long view directs attention away from individuals and toward trends. This chapter redirects the book's focus to individuals. We are not entirely powerless in respect of technological trends. There are steps we can take to humanize the Digital Age. We can adjust our attitudes and responses to digital technologies and the corporations marketing them. This chapter explored five ways to approach the Digital Age with an empowered uncertainty.

9 Making a Very Human Digital Age

Chapter 1 of this book presented a challenge to human agency in the Digital Age. Digital machines are increasingly doing mind work better and more cheaply than we are. The fast trajectory of AI's improvement suggests that this colonization of the domain of mind work will accelerate.

It is important to understand that this is a threat not to all instances of work but instead to the work norm. It is a threat to the idea that it is normal for humans to leave school and find work. Perhaps the specific combination of human insights that make for good stand-up comedy can never be replicated by a robot, no matter how massive and well-analyzed its database of Richard Pryor and Joan Rivers sketches. The existence of a few stand-up comedians or therapeutic masseurs in Digital Age societies does not suffice to sustain the work norm. We will need sufficiently many such jobs to justify our children's expectations that they will reach maturity and find satisfying ways to contribute to their society's collective well-being and receive, in return, a fair share of that society's wealth.

In chapter 4 I warned against overconfident predictions about the Digital Age. It's possible that current anxieties about the consequences of automation will turn out to be mistaken. The next stage of the Digital Revolution may see the creation of many new jobs that engage human imaginations and agency in ways impossible for people in the first decades of the twenty-first century to imagine. Perhaps our grandchildren, doing thrilling Digital Age jobs, will pity us the soulless drudgery that we call work and wonder how we put up with it for so long. I sincerely hope that this is the case. However, we should distinguish our hopes for the Digital Age from our rational expectations. In this book, I suggest that we approach the uncertainties of the future with an insurance mindset. We can hope for the human-friendliest version of the Digital Age but still prepare for a future in which the Digital Revolution poses a genuine threat to the work norm.

Perhaps the thinking that we do now about how to create jobs likely to survive significant technological progress will not be required for the Digital Revolution. But it may be useful for the technological revolution that comes after that. It may give our descendants some peace of mind as they confront the opportunities and disruptions of a Quantum Revolution that ejects from the economy many humans who work with digital technologies, replacing them with the select few humans who really understand how quantum mechanical technologies function. The insurance mindset suggests that the cost of such precautions is small. They require some creative thinking about possible futures for human beings—creative thinking that goes beyond the version of the future presented by technologists as inevitable.

Suppose we take seriously the threat to the work norm from the Digital Revolution. The question is how to respond. I presented an ideal—the ideal of the social-digital economy. We should permit and indeed encourage the colonization of certain domains of mind work by machines. But we should insist on the protection and enlargement of essentially social mind work. How do we decide which jobs to protect and which to surrender to the machines? The jobs that we seek to protect are those centered on interactions between human minds. It matters to us that there is a human mind with thoughts and feelings much like ours behind a performance of Hamlet or advice about how to confront a depressive illness. When we reflect, we find that we care about the connection that the job establishes between us and other human minds.

No objective fact about a task directs us to value it purely in terms of efficiency or partly in terms of humanness. In chapter 6 I suggested that our limited contact with the pilots who fly our jetliners means that we will not mourn the arrival of more efficient fully-automated cockpits. But it didn't have to be this way. Suppose that the need to protect against hijackers had not led to an increasing sequestration of pilots. Suppose people arriving after a long trip in a jetliner routinely reflected that one of the most enjoyable parts of their journey was getting to chat with the pilot and "helping" to fly the plane. We might have judged that the enjoyment of our contacts with human pilots justified a small increase in the probability of a crash. People who enjoy skiing freely indulge a pastime that they should acknowledge as raising their probability of violent death. So it could have been for people who choose human piloted jetliners in a future when the pilotless versions are manifestly safer. But that doesn't seem to be the way we have gone. The inauguration of the era of pilotless passenger jets will occur without a great impact on the experience of passengers. It may be a shock for the first passengers of pilotless planes to learn that "there's no one flying this thing!" But reports on how quickly people get used to being driven around by driverless cars suggest that we will rapidly adapt. We will quickly come to celebrate the increases in safety that come with fully-automated cockpits.

Can we imagine the full range of social jobs that we could create were we sufficiently enthused by the ideal of a social-digital economy? Probably not. In chapters 5 and 6 I offered disparate examples of jobs centered on connections with other human minds. I suggested that the value of these connections more than compensates for losses in efficiency. The Digital Age could feature human baristas, personal shoppers, actors, and space explorers. These jobs alone will not carry the work norm into the Digital Age. Fortunately, it's reasonable for us to expect that this quite disparate assortment of jobs centered on connections between human minds will give rise to a much more diverse collection of essentially social jobs.

In chapter 4, I criticized David Autor's confidence that the Digital Revolution will bring jobs that we today cannot imagine. He says that we should expect the economic growth resulting from technological progress to generate new, previously unimaginable jobs. In response to Autor, I granted that technological progress reliably produces new economic roles. There is, however, the further question about who or what fills those roles. I proposed that machine learning has the potential to cheaply and efficiently fill many of the novel economic roles created by the Digital Revolution. The insurance mindset suggests that we must take seriously the possibility that few of the new economic roles created by the Digital Revolution will be filled by human workers.

In what follows I appeal to reasoning similar to that offered by Autor to suggest that future social economies could contain a great variety of jobs. We can infer the existence of these jobs even when we fail to imagine their details. When we try to describe them, they may seem somewhat bizarre and fantastical—just how many socially enhanced personal shoppers will the social economy of the Digital Age need? But these quite disparate examples suggest many jobs based on our social needs and capacities. I can offer a more secure basis for the existence of currently unimagined future jobs than that offered by Autor. The new jobs of the social economy will form around human social capacities and needs that are left comparatively untouched by technological change. Many of these social needs are currently unsatisfied in our age's most prosperous and technologically advanced societies.

In chapter 5 I commented on the epidemic of social isolation that seems to afflict today's most technologically advanced societies. I described the work of John Cacioppo on our "obligatorily gregarious" natures.[1] The fact that we tend to suffer when isolated is the legacy of our evolution as a gregarious species. This history suggests possible locations for many social economy jobs. We can look back at the history of foraging common to all humans alive today and consider the diverse ways in which our ancestors satisfied each other's social needs. Each of these suggests the possibility for jobs capable of surviving the transition to the Digital Age. Our bias toward humans leads us to always find machines inadequate in these roles.

This suggests a question. Foragers can be busy, but they do not have jobs. Why must our social needs be met by the seemingly clumsy maneuver of inventing jobs to meet them? Advocates of the UBI warn against the conflation of work with jobs. Mesolithic foragers engaged in the kind of purposive activity characteristic of work. But they did not have jobs. Their work was not done in expectation of a salary. Forager communities feature work but not the institution of work.

Work can function as the social glue of complex and diverse societies of the Digital Age. We evolved from what Paul Seabright calls a "shy, murderous ape that had avoided strangers throughout its evolutionary history."[2] The expansive diversity of most technologically advanced societies makes use of the institution of work to bind people together. When we work together we overcome, to some extent, boundaries of race, religion, gender, and ability. A member of a formerly despised group becomes your colleague or valued customer. As we saw in chapter 5, there is support for the idea that working together to address a challenging goal is an excellent way to create trust. Work provides social glue that helps to fashion strangers into cohesive mutually trusting societies.

This is not to say that work is the only way in which we can overcome these boundaries. Sport is another context in which we must work with others to succeed. You may have been raised by your parents to feel suspicious about people from Asia, but when your weekend football team's leading scorer is Asian you find that childhood belief challenged. But no collaborative context matches the contemporary significance of the institution of work. If advocates of the UBI are to significantly diminish the social significance of work, they must do much more than rely on the mere theoretical possibility of institutions in the jobless societies of the Digital Age that match the societal significance of work. We should be alert to the danger that society will fracture into less expansive societies defined around ethnicity, race, or religion. The work norm is a large part of the way we maintain liberal democratic diversity. How, given the parochial nature of our obligatory gregariousness, can future societies with the UBI not fragment into social units of a few hundred, a typical size of forager communities? Work coaxes us into contact and collaboration with people who differ from us in ways that many of us are raised to find significant. I grant that there may be solutions to this problem that do not require the institution of work. But we cannot just wish these solutions into existence. It would be a mistake to forsake work for something that might, in theory, do the same job of forming strangers into cohesive societies. The institution of work and the work norm is our current solution. It works!

When I endorse the work norm I do not endorse many of the forms that work takes in this economically uncertain time. Much of today's work is unsatisfactory. We have an inconsistent attitude toward work done by the poor and work done by the rich. Our attitude toward work done by the poor is well summarized by the economist's view of work as generating negative personal utility. Your job is something that you would not do if you were not paid. You justify turning up to work because you expect that the positive personal utility you will derive from your salary will more than compensate for the negative utility of doing your job. Employers seek to pay their workers enough to make it worth their while turning up to work, but no more than that. This is not how we think about work done by the wealthy. According to *Vanity Fair* magazine, Matt Damon may have received payment for his performance in the 2016 movie, *The Bourne Ultimatum*, that worked out at over $1 million per line of dialogue.[3] We do not look at that figure and speculate about the intense torment caused by speaking each of those lines that required massive sums of money to offset it. We expect Damon to tell us that he enjoyed acting in the movie.

I hope that the social-digital economy will enable us to generalize Damon's experience of work. Damon justifies his salary by pointing to the magnitude of the good that his acting generates. People like his movies and are prepared to pay to see them. The jobs of the social economy require us to interact with other people. The fact that socializing is work doesn't preclude our enjoyment of it. Salaries in the social economy will be justified not in terms of compensating for the negative personal utility of work, but instead in terms of the magnitude of the benefits we bring to society.

The institution of work is not perfect as it is currently realized in early twenty-first century liberal democracies. We should always be looking for ways to eliminate its injustices. The ideal of a social economy in which the social work we do for strangers is properly valued offers a way forward.

Welcoming a Social Age

There is a campaign to rename our current geological epoch. According to convention in geology, we are in the Holocene Epoch—a name formed out of the Ancient Greek words for wholly new. The Holocene dates from about 12,000 years ago. Some geologists think we should acknowledge that we have passed into a new epoch—the Anthropocene Epoch, a name that indicates the significance of human impacts on the planet's geology and ecosystems. I would like to use these final pages to start a campaign for a change in the way we name eras in human history. This change would be contingent on our taking serious steps to implement a social-digital economy. In the introduction, I offered a famous quote from Francis Bacon that presented technological change as the driver of human history. According to Bacon, "no empire, no sect, no star seems to have exerted greater power and influence in human affairs" than the biggest technological advances. We name our historical eras for their dominant technologies. We progressed through the Stone Age, the Bronze Age, and the Iron Age. The Industrial Age was named for the technologies of the Industrial Revolution. And so, we arrive at the Digital Age, an age defined by the technological package of the networked digital computer.

Suppose we act on the advice offered in this book. It would be appropriate to name the age we will enter as the Social Age, named for the significance to human affairs of social interactions. We would reject the suggestion that "social" be coopted as the name for a very profitable category of digital technologies—the social networking technologies of Facebook, Twitter, LinkedIn, and so on. Rather it would indicate the significance to our collective human experience of relationships between human minds.

The choice of the name Social Age departs from the convention of naming ages for our dominant technologies. But it represents no rejection of technology. The citizens of the Social Age would not live in some technology-free idyll, magically harmonized with nature. They will benefit from very powerful digital technologies. They will acknowledge that digital technologies are essential to the lives they lead. But they will nevertheless view these essential technologies as of secondary significance to their understanding of the lives they lead. We have seen this downgrading of the importance of older yet still important technologies in our own time. The lives led by most humans today would be impossible without electricity. Moreover, without electricity, there is no Google. But we do not think of Google as an electricity company where the modifier "electricity" indicates not that it is involved in the generation of electricity, but that its every activity requires electricity, the recognition that without electricity, there are no Google searches and no application of AI to the results of these searches. Electricity has obligingly receded into the background. We give most credit for Google's achievements to the newer, "game changing" digital technologies. Digital technologies will be essential for much that happens over the next decades. But we may come to recognize that relationships between human minds nevertheless have a greater significance to who we are both individually and collectively. We could move on from the Digital Age to the Social Age in recognition of this general change in attitude. The Social Age might eventually give way to an age whose human fundamentals are different. But this is unlikely to happen because of a change in our technologies. We would not look to the introduction of a new technological package to end the Social Age.

The name "Anthropocene Epoch" suggests a negative evaluation of human accomplishments. We are a principally negative force interfering with the benign natural processes that defined the Holocene. We cause mass extinctions, pollute the seas, and emit lots of CO_2. The choice of the name "Social Age" for a coming age in human history indicates a more positive evaluation of what we are jointly capable of. I do not offer the Social Age as a prediction about the next key large scale development in human history. We may choose not to realize the ideal of the social-digital economy and so continue to view our technologies as the principle influences over collective human experience. It's important that we acknowledge the long-term implications of rejecting this opportunity to make our existences more social. Suppose that we reject the suggestion that we preserve a special significance for connections between human minds. We should fear a dehumanized future thoroughly dominated by the value of efficiency. This is effectively a voluntary path to extinction in which we choose to cede our places to our robotic betters.

Notes

Introduction: Taking the Long View of Digital Revolution

1. Rose-Mary Sargent (ed.), *Francis Bacon: Selected Philosophical Works* (Indianapolis: Hackett, 1999), 146. See Erik Brynjolfsson and Andrew McAfee, *The Second Machine Age: Work, Progress, and Prosperity in a Time of Brilliant Technologies* (New York: W. W. Norton, 2014) for a modern presentation of the significance of technological change to human history.

2. See Walter Isaacson, *The Innovators: How a Group of Hackers, Geniuses, and Geeks Created the Digital Revolution* (New York: Simon and Schuster, 2014) for an illuminating history of the Digital Revolution's technologies and personalities.

3. The term "technological unemployment" was introduced by John Maynard Keynes, "Economic Possibilities for Our Grandchildren," in his *Essays in Persuasion* (New York: W. W. Norton & Co., 1963).

4. "Will a Robot Take Your Job?" *BBC News*, September 11, 2015, http://www.bbc.com/news/technology-34066941.

5. Rory Cellan-Jones, "Stephen Hawking Warns Artificial Intelligence Could End Mankind," *BBC News*, December 2, 2014, http://www.bbc.com/news/technology-30290540; Nick Bostrom, *Superintelligence: Paths, Dangers, Strategies* (Oxford: Oxford University Press, 2014).

6. For my response to fears about artificial superintelligence, see Nicholas Agar, "Don't Worry about Superintelligence," *Journal of Evolution and Technology* 26: 1 (2016): 73–82. Bostrom overstates the threat from a human-unfriendly superintelligence. We shouldn't worry too much about a small threat of extinction from AI just as we don't worry much about the threat of personal extinction when we carefully cross a busy road.

7. Samuel Gibbs, "Apple Co-founder Steve Wozniak Says Humans Will Be Robots' Pets," *Guardian*, June 25, 2015, https://www.theguardian.com/technology/2015/jun/25/apple-co-founder-steve-wozniak-says-humans-will-be-robots-pets.

8. Elon Musk raised the prospect of humanity as pets for future AIs. According to Musk, the only way to avoid futures as house cats was to become cyborgs. Since we can't beat machines of the Digital Age, we must then become them, at least partially. See James Tibcomb, "Elon Musk: Become cyborgs or risk humans being turned into robots' pets," *Telegraph,* June 2, 2016, http://www.telegraph.co.uk/technology/2016/06/02/elon-musk-become-cyborgs-or-risk-humans-being-turned-into-robots.

9. Stanley Coren, "How Many Dogs Are There In the World?" *Psychology Today,* September 19, 2012, https://www.psychologytoday.com/blog/canine-corner/201209/how-many-dogs-are-there-in-the-world.

10. Nick Bilton, "How the Media Screwed Up the Fatal Tesla Accident," *Vanity Fair,* July 7, 2016, https://www.vanityfair.com/news/2016/07/how-the-media-screwed-up-the-fatal-tesla-accident.

11. Daniel Wegner and Kurt Gray, *The Mind Club: Who Thinks, What Feels, and Why It Matters* (New York: Penguin, 2016), 3.

12. John Cacioppo and William Patrick, *Loneliness: Human Nature and the Need for Social Connection* (New York: W. W. Norton & Company, 2008; Kindle).

13. Ibid., loc. 928–929.

14. Ibid., loc. 931.

15. Robert Putman, *Bowling Alone: The Collapse and Revival of American Community* (New York: Simon & Schuster, 2000), 213.

16. Martin Ford, *Rise of the Robots: Technology and the Threat of a Jobless Future* (New York: Basic Books, 2015).

17. Robert Gordon, *The Rise and Fall of American Growth: The U.S. Standard of Living Since the Civil War* (Princeton, NJ: Princeton University Press, 2016; Kindle), loc. 6260.

18. See, for example, Joris Toonders, "Data Is the New Oil of the Digital Economy," *Wired,* July 2014, https://www.wired.com/insights/2014/07/data-new-oil-digital-economy. For skepticism about this claim, see Jer Thorp, "Big Data Is Not the New Oil," *Harvard Business Review*, November 30, 2012, https://hbr.org/2012/11/data-humans-and-the-new-oil.

Chapter 1: Is the Digital Revolution the Next Big Thing?

1. V. Gordon Childe, *The Dawn of European Civilisation* (London: Kegan Paul, 1925); V. Gordon Childe, *The Most Ancient Near East: The Oriental Prelude to European Prehistory* (London: Kegan Paul, 1928); V. Gordon Childe, *Man Makes Himself* (London:

Coronet, 2003). See also Graeme Barker, *The Agricultural Revolution in Prehistory: Why Did Foragers Become Farmers?* (Oxford: Oxford University Press, 2006), chap. 9; and Steven Mithen, *After the Ice: A Global Human History* (London: Weidenfeld and Nicolson, 2003), chap. 7 and 21.

2. See William Rosen's discussion of technological hubs. Rosen argues for the salience of the steam engine, identifying it as the hub of the technologies introduced by the Industrial Revolution. "Its central position connecting the era's technological and economic innovations: the hubs through which the spokes of coal, iron, and cotton were linked." William Rosen, *The Most Powerful Idea in the World: A Story of Steam, Industry, and Invention* (Chicago: University of Chicago Press, 2010), xxi.

3. Steven Mithen, *After the Ice: A Global Human History* (London: Weidenfeld and Nicolson, 2003; Kindle), loc. 1383.

4. Ibid., chap. 7.

5. Robert Gordon, *The Rise and Fall of American Growth: The U.S. Standard of Living Since the Civil War* (Princeton, NJ: Princeton University Press, 2016; Kindle).

6. Ibid., loc. 174.

7. Ibid., loc. 6253.

8. Ibid., loc. 255.

9. Ibid., loc. 11031.

10. Ibid., loc. 6260.

11. Ibid., loc. 8233–8234.

12. Ibid., loc. 172

13. Ibid., chap.18.

14. Ibid., loc. 10294.

15. Ray Kurzweil is the most influential recent booster of exponential technological progress. See Ray Kurzweil, *The Singularity Is Near: When Humans Transcend Biology* (New York: Viking, 2005).

16. Erik Brynjolfsson and Andrew McAfee, *The Second Machine Age: Work, Progress, and Prosperity in a Time of Brilliant Machines*, (New York: W. W. Norton, 2014).

17. Thomas Friedman, *Thank You for Being Late; Finding a Job, Running a Country, and Keeping Your Head in an Age of Accelerations* (New York: Farrar, Straus and Giroux, 2016).

18. Gordon, *The Rise and Fall of American Growth*, Kindle loc. 6263.

19. Lauren Johnson, "Google's Ad Revenue Hits $19 Billion, Even as Mobile Continues to Pose Challenges," *Adweek*, July 28, 2016, http://www.adweek.com/?p=172722.

20. Tim Wu, *The Attention Merchants: The Epic Scramble to Get Inside Our Heads* (New York: Knopf, 2016).

21. See the informative review of Gordon's book by Tyler Cowen. Cowen summarizes his response: "In a nutshell, Gordon is probably right about the past, but wrong about the future." Tyler Cowen, "Is Innovation Over? The Case Against Pessimism," *Foreign Affairs,* March/April 2016, https://foreignaffairs.com/reviews/review-essay/2016-02-15/innovation-over.

22. Quoted in Ashlee Vance, "This Tech Bubble Is Different," *Bloomberg,* April 15, 2011, https://www.bloomberg.com/news/articles/2011-04-14/this-tech-bubble-is-different.

23. Christof Koch, "How the Computer Beat the Go Master," *Scientific American,* March 19, 2016, https://www.scientificamerican.com/article/how-the-computer-beat-the-go-master.

24. Cited in World Health Organization, Fact Sheet no. 310, "The Top Ten Causes of Death," updated May 2014, http://www.who.int/mediacentre/factsheets/fs310/en/.

25. "Number of casualties due to terrorism worldwide between 2006 and 2016," Statista, 2018, https://www.statista.com/statistics/202871/number-of-fatalities-by-terrorist-attacks-worldwide.

26. "Tesla driver killed while using autopilot was watching *Harry Potter,* witness says," *Guardian,* July 1, 2016, https://www.theguardian.com/technology/2016/jul/01/tesla-driver-killed-autopilot-self-driving-car-harry-potter.

27. Hod Lipson and Melba Kurman, *Driverless: Intelligent Cars and the Road Ahead* (Cambridge, MA: MIT Press, 2016; Kindle), loc. 4167–4171.

28. Ibid., loc. 420.

29. Gordon, *The Rise and Fall of American Growth,* loc. 11481–11485.

30. Kevin Rawlinson, "Fewer car owners and more driverless vehicles in future, survey reveals," *Guardian,* January 9, 2017 https://www.theguardian.com/business/2017/jan/09/fewer-car-owners-more-driverless-vehicles-future-survey-reveals.

31. Gordon, *The Rise and Fall of American Growth,* loc. 11490.

32. For an account of potential economic benefits of driverless technology more informative than Gordon's imaginative failure, see chapter 2 of Lipson and Kurman, *Driverless.*

33. Gordon, *The Rise and Fall of American Growth,* loc. 11479.

34. David Autor, "Why Are There Still So Many Jobs? The History and Future of Workplace Automation," *Journal of Economic Perspectives* 29 (3) 2015: 3–30, at 25–26.

35. Emily Retter, "The Stupidest Quiz Show Answers EVER," *Mirror*, January 22, 2016, https://www.mirror.co.uk/tv/tv-news/stupidest-quiz-show-answers-ever-7229447.

36. Autor, "Why Are There Still So Many Jobs?," 26.

37. Ibid.

38. Gerd Gigerenzer, *Gut Feelings: The Intelligence of the Unconscious* (New York: Viking, 2007), 85.

39. Erika Check Hayden, "The Rise and Fall and Rise Again of 23andMe," *Nature*, October 11, 2017, https://www.nature.com/news/the-rise-and-fall-and-rise-again-of-23andme-1.22801.

40. Daniela Hernandez, "Big Tech Has Your Email and Photos. Now It's on a Quest to Own Your DNA," *Huffington Post*, July 20, 2015, https://www.huffingtonpost.com/entry/big-tech-dna_55ac3376e4b0d2ded39f46eb?utm_hp_ref=worldpost-future-series.

41. Pedro Domingos, *The Master Algorithm: How the Quest for the Ultimate Learning Machine Will Remake Our World* (London: Allen Lane, 2015), xvii.

42. Siddhartha Mukherjee, *The Emperor of All Maladies: A Biography of Cancer* (New York: Scribner, 2013; Kindle); loc. 9792–9794.

43. Pedro Domingos, *The Master Algorithm: How the Quest for the Ultimate Learning Machine Will Remake Our World* (London: AllenLane, 2015), 259–261.

44. Ibid, 259.

45. Ibid, 259.

46. Olivia Solon, "Mark Zuckerberg and Priscilla Chan aim to 'cure, prevent and manage' all disease," *Guardian*, September 21, 2016, https://www.theguardian.com/technology/2016/sep/21/mark-zuckerberg-priscilla-chan-end-disease.

Chapter 2: AI's Split Personality—Minds or Mind Workers?

1. Alan Turing, "Computing Machinery and Intelligence," *Mind* 49 (1950): 433–460.

2. For excellent philosophical presentations of Turing and his test, see Jack Copeland, *Artificial Intelligence: A Philosophical Introduction* (Oxford: Blackwell, 1993); Jack Copeland, "The Turing Test," *Minds and Machines* 10 (2000): 519–539; Daniel Dennett, "Can Machines Think?" in Michael Shafrom (ed.), *How We Know* (San Francisco: Harper and Row, 1985), 121–145; Hector Levesque, *Common Sense, the Turing Test, and*

the Quest for Real AI (Cambridge, MA: MIT Press, 2017; Kindle.); James Moor, "An Analysis of the Turing test" *Philosophical Studies* 30 (1976): 249–257; Graham Oppy and David Dowe, "The Turing Test," *Stanford Encyclopedia of Philosophy*, https://plato.stanford.edu/entries/turing-test.

3. See the discussion in Levesque, *Common Sense, the Turing Test, and the Quest for Real AI*. Levesque distinguishes between adaptive machine learners capable of learning from large amounts of data in an unsupervised way and the goal of GOFAI (Good Old Fashioned Artificial Intelligence), which is to make a machine with common sense, the kind of intelligence that humans apply to everyday life. We can identify GOFAI with the philosophical interest in AI. Adaptive machine learning corresponds to the pragmatic interest.

4. Most famously John Searle, "Minds, Brains, and Programs," *Behavioral and Brain Sciences* 3 (1980): 417–424.

5. Pedro Domingos, *The Master Algorithm: How the Quest for the Ultimate Learning Machine Will Remake Our World* (London: Allen Lane, 2015), xvii.

6. Ibid., 25.

7. Quoted in Andrew Hodges, *Alan Turing: The Enigma*, Centenary Edition (Princeton: Princeton University Press, 2013), 251.

8. Searle, "Minds, Brains, and Programs."

9. Douglas Hofstadter's 1979 book *Gödel, Escher, Bach* attracted widespread excitement about the challenge of programming a machine to think. Recently Hofstadter has become disenchanted by artificial intelligence and what he sees as its loss of interest in human thought. Hofstadter expresses his disappointment about Deep Blue's 1997 victory over Kasparov that we can recognize as coming from his philosophical interest in AI. He says, "Deep Blue plays very good chess—so what? Does that tell you something about how we play chess? No. Does it tell you about how Kasparov envisions, understands a chessboard?" (James Somers, "The Man Who Would Teach Machines to Think," *Atlantic*, November 2013, https://www.theatlantic.com/magazine/archive/2013/11/the-man-who-would-teach-machines-to-think/309529.) Again, Hofstadter's answer is presumably no. I suspect that these queries are unlikely to lead IBM to regret its thoroughgoing pragmatism about machine chess.

10. Turing, "Computing Machinery and Intelligence," 442.

11. Most commentators hold the view presented by Graham Oppy and David Dowe: "There is little doubt that Turing would have been disappointed by the state of play at the end of the twentieth century." One exception is Jack Copeland, "The Turing Test," *Minds and Machines*, 10 (2000): 519–539.

12. Turing, "Computing Machinery and Intelligence," 442.

13. To chat with Sgt. Star, the US Army's virtual guide, go to https://www.goarmy.com/ask-sgt-star.html.

14. Consider the negative assessment of the Loebner Prize by Marvin Minsky, widely acclaimed as the father of artificial intelligence. Minsky called the Loebner Prize "obnoxious and stupid." He "offered a cash award of his own to anybody who can persuade Loebner to abolish his prize and go back to minding his own business." John Sundman, "Artificial Stupidity," *Salon*, February 27, 2003, https://www.salon.com/2003/02/26/loebner_part_one.

15. Brian Christian, *The Most Human Human: What Artificial Intelligence Teaches Us About Being Alive* (New York: Anchor Books, 2011) discusses his participation in the Loebner Prize.

16. Levesque, *Common Sense, the Turing Test, and the Quest for Real AI*, loc. 736.

17. To chat with ELIZA, computer therapist, visit http://www.manifestation.com/neurotoys/eliza.php3.

18. David Morris, "Ashley Madison Used Chatbots to Lure Cheaters, Then Threatened to Expose Them When They Complained," *Fortune*, July 10, 2016, http://fortune.com/2016/07/10/ashley-madison-chatbots.

19. Nicholas Agar, "Reflections on 'Chatbot,'" *OUPblog*, November 25, 2016, https://blog.oup.com/2016/11/reflections-on-chatbot-woty-2016.

20. For an appeal to our hyperactive agency detection device to explain belief in God, see Justin Barrett, *Why Would Anyone Believe in God?* (Walnut Creek, CA: AltaMira Press, 2004).

21. See the Wikipedia entry on Deep Blue at https://en.wikipedia.org/wiki/Deep_Blue_(chess_computer).

Chapter 3: Data as a New Form of Wealth

1. For a discussion of the long-term influence of technological progress on human well-being see Nicholas Agar, *The Sceptical Optimist: Why Technology Isn't the Answer* (Oxford: Oxford University Press, 2015).

2. "Facebook Valuation Tops $200 Billion," *Bloomberg*, September 8, 2014, https://www.bloomberg.com/infographics/2014-09-08/facebook-valuation-tops-200-billion.html.

3. Patrick Gillespie, "Apple: First U.S. company worth $700 billion," *CNN Money*, February 10, 2015, http://money.cnn.com/2015/02/10/investing/apple-stock-high-700-billion.

4. Bezos's net worth in early 2018 was US $132.1 billion. See "The World's Billionaires," *Forbes*, updated daily, http://www.forbes.com/billionaires/list. For a discussion of Bezos's holdings in real estate, see Nav Athwal, "The Fabulous Real Estate Portfolio of Jeff Bezos," *Forbes,* August 13, 2015, https://www.forbes.com/sites/navathwal/2015/08/13/the-fabulous-real-estate-portfolio-of-jeff-bezos/#90475c0569cd.

5. See, for example, Joris Toonders, "Data Is the New Oil of the Digital Economy," *Wired,* July 2014, https://www.wired.com/insights/2014/07/data-new-oil-digital-economy. For skepticism, see Jer Thorp, "Big Data Is Not the New Oil," *Harvard Business Review*, November 30, 2012, https://hbr.org/2012/11/data-humans-and-the-new-oil.

6. For an analysis that locates Google's leadership in Search more in the quantities of its data than in the cleverness of its algorithms, see Pedro Domingos, *The Master Algorithm: How the Quest for the Ultimate Learning Machine Will Remake Our World* (London: Allen Lane, 2015), 12.

7. "Data Is the New Oil, Analytics the New Refinery," Retail Info Systems, March 30, 2015, https://risnews.edgl.com/retail-news/Data-is-the-New-Oil,-Analytics-the-New-Refinery99333.

8. "The World's Billionaires," *Forbes*, updated daily, http://www.forbes.com/billionaires/list.

9. Charles Riley, "Mark Zuckerberg Gives Pope Francis a Drone," *CNN,* August 29, 2016, http://money.cnn.com/2016/08/29/technology/pope-francis-mark-zuckerberg-facebook-italy.

10. For an interesting critique of the "great man of tech myth," see Amanda Schaffer, "Tech's Enduring Great Man Myth," *MIT Technology Review,* August 4, 2015, https://www.technologyreview.com/s/539861/techs-enduring-great-man-myth.

11. Jaron Lanier, *Who Owns the Future?* (New York: Simon and Schuster, 2013), 79.

12. Naina Bajekal, "Londoners Unwittingly Exchange First Born Children For Free Wi-Fi," *Time*, September 29, 2014, http://time.com/3445092/free-wifi-first-born-children.

13. "Power of One Million," *23andMe Blog*, June 18, 2015, https://blog.23andme.com/news/one-in-a-million.

14. Daniela Hernandez, "Big Tech Has Your Email and Photos. Now It's on a Quest to Own Your DNA," *Huffington Post*, July 20, 2015, https://www.huffingtonpost.com/entry/big-tech-dna_55ac3376e4b0d2ded39f46eb.

15. Dorothy Denning, "Concerning Hackers Who Break into Computer Systems," *Proceedings of the 13th National Computer Security Conference* (Washington, DC: National Institute of Standards and Technology/National Computer Security Center, 1990), 653–664.

16. Thomas Jefferson, "Thomas Jefferson to Isaac McPherson, 13 Aug. 1813," in *Founders' Constitution*, eds. Philip B. Kurland and Ralph Lerner (Indianapolis: Liberty Fund, 1986).

17. Jeremy Rifkin, *The Zero Marginal Cost Society: The Internet of Things, the Collaborative Commons, and the Eclipse of Capitalism* (New York: Palgrave MacMillan, 2014).

18. Ibid, 18.

19. Kevin Kelly, *The Inevitable: Understanding the 12 Technological Forces That Will Shape Our Future* (New York: Penguin, 2016).

20. Cory Doctorow, *Information Doesn't Want to Be Free: Laws for the Internet Age* (San Francisco: McSweeney's, 2014), section 1.4.

21. For an illuminating account of the reality of holding power, see Bruce Bueno de Mesquita and Alastair Smith, *The Dictator's Handbook: Why Bad Behavior Is Almost Always Good Politics* (New York: Public Affairs, 2011).

22. Jim McLauchlin, "*Star Wars'* $4 Billion Price Tag Was the Deal of the Century," *Wired,* December 14, 2015, https://www.wired.com/2015/12/disney-star-wars-return-on-investment.

23. Kelly, *The Inevitable*, chap. 3.

24. Alex Hern, "Facebook Is Making More and More Money from You. Should You Be Paid for It?" *Guardian,* September 25, 2015, https://www.theguardian.com/technology/2015/sep/25/facebook-money-advertising-revenue-should-you-be-paid.

25. See *Statista: The Statistics Portal*, https://www.statista.com/statistics/264810/number-of-monthly-active-facebook-users-worldwide.

26. Antonio Garcia Martinez, *Chaos Monkeys: Inside the Silicon Valley Money Machine* (New York: Harper, 2016).

27. Lanier, *Who Owns the Future?*, 40.

28. See the Wikipedia page on Lanier, https://en.wikipedia.org/wiki/Jaron_Lanier.

29. Lanier, *Who Owns the Future?*, 9.

30. Ibid., 216–220.

31. Ibid., 218.

32. Ibid., 263.

33. Ibid., 5.

34. Theo Valich, "Facebook Bank: Introduces Micro Payments up to $10,000," *VR World,* June 27, 2015, http://vrworld.com/2015/06/27/facebook-bank-introduces-micro-payments-up-to-10000.

35. James Fair, "Hunting Success Rates: How Predators Compare," *Discover Wildlife*, December 17, 2015, http://www.discoverwildlife.com/animals/hunting-success-rates-how-predators-compare.

36. Bruce Schneier, *Data and Goliath: The Hidden Battles to Collect Your Data and Control Your World* (New York: W.W. Norton, 2015), 17.

Chapter 4: Can Work Be a Norm for Humans in the Digital Age?

1. Thomas Leary, "Industrial Ecology and the Labor Process," in Charles Stephenson and Robert Asher (eds.), *Life and Labor: Dimensions of American Working-Class History* (New York: SUNY Press, 1986), 44.

2. Robert Gordon, *The Rise and Fall of American Growth: The U.S. Standard of Living Since the Civil War* (Princeton, NJ: Princeton University Press, 2016; Kindle), loc. 4989.

3. Mihaly Csikszentmihalyi and Judith LeFevre, "Optimal Experience in Work and Leisure," *Journal of Personality and Social Psychology* 56 (1989): 815–822.

4. Mihaly Csikszentmihalyi, *Flow: The Psychology of Happiness*, revised and updated edition (London: Random House, 2002).

5. Csikszentmihalyi and LeFevre, "Optimal Experience in Work and Leisure."

6. Kevin Kelly, *The Inevitable: Understanding the 12 Technological Forces That Will Shape Our Future* (New York: Viking, 2016), 50.

7. John Maynard Keynes, "Economic Possibilities for Our Grandchildren," *Essays in Persuasion* (New York: W. W. Norton and Co., 1963), 358–373.

8. See also David Autor, "Why Are There Still So Many Jobs? The History and Future of Workplace Automation," *Journal of Economic Perspectives* 29, no. 3 (2015): 3–30.

9. Kelly, *The Inevitable*, 60.

10. Autor, "Why Are There Still So Many Jobs?"

11. Ibid., 6.

12. Andrew McAfee and Erik Brynjolfsson, *Machine, Platform, Crowd: Harnessing Our Digital Future* (New York: W. W. Norton and Company, 2017), 123.

13. David Autor and David Dorn, "The Growth of Low-Skill Service Jobs and the Polarization of the US Labor Market," *American Economic Review* 103, no. 5 (2013): 1553–1597.

14. Autor and Dorn, "The Growth of Low-Skill Service Jobs and the Polarization of the US Labor Market," 1555.

15. David Dorn, "The Rise of the Machines: How Computers Have Changed Work," *UBS International Center of Economics in Society*, Public Paper #4, December 16, 2015, 14, http://www.zora.uzh.ch/id/eprint/116935.

16. See Gordon's discussion of the features of delivery truck driving jobs that make them difficult to automate. He suggests that driverless trucks are, at best, a very partial solution. For Gordon, the problem arises when the trucks arrive at the destinations of their deliveries. He observes, "it is remarkable in this late phase of the computer revolution that almost all placement of individual product cans, bottles, and tubes on retail shelves is achieved today by humans rather than robots. Driverless delivery trucks will not save labor unless work is reorganized so that unloading and placement of goods from the driverless trucks is taken over by workers at the destination location" (Gordon, *The Rise and Fall of American Growth,* Kindle loc. 11495).

17. David Dorn, "The Rise of the Machines: How Computers have Changed Work," *UBS International Center of Economics in Society*, Public Paper #4, December 16, 2015, 16, http://www.zora.uzh.ch/id/eprint/116935.

18. Marcus Wohlsen, "A Rare Peek Inside Amazon's Massive Wish-Fulfilling Machine," *Wired*, June 16, 2014, https://www.wired.com/2014/06/inside-amazon-warehouse.

19. Pedro Domingos, *The Master Algorithm: How the Quest for the Ultimate Learning Machine Will Remake Our World* (London: Allen Lane, 2015), 12.

20. Andrew McAfee and Erik Brynjolfsson, *Machine, Platform, Crowd: Harnessing Our Digital Future* (New York: W. W. Norton and Company, 2017), 123.

21. Mark Zuckerberg, Facebook post, January 3, 2016, https://www.facebook.com/zuck/posts/10102577175875681.

22. For exploration of how the idea of moral insurance applies to some controversial claims of utilitarians, see Nicholas Agar, "How to Insure against Utilitarian Overconfidence," *Monash Bioethics Review* 32 (2014): 162–171.

23. Among modern authors who inclined to mock the sentiment expressed by the *Times* are Steven Levitt and Stephen Dubner, *SuperFreakonomics: Global Cooling, Patriotic Prostitutes, and Why Suicide Bombers Should Buy Life Insurance* (New York: William Morrow, 2009), introduction.

24. Ben Johnson, "The Great Horse Manure Crisis of 1894," *Historic UK*, http://www.historic-uk.com/HistoryUK/HistoryofBritain/Great-Horse-Manure-Crisis-of-1894.

25. This widely cited *Times of London* quote has an interesting history. See Rose Wild, "We Were Buried in Fake News as Long Ago as 1894," *Sunday Times*, January 13, 2018, https://www.thetimes.co.uk/article/we-were-buried-in-fake-news-as-long-ago-as-1894-ntr23ljd5.

Chapter 5: Caring about the Feelings of Lovers and Baristas

1. See Nicholas Agar, "Let's Treat Robots Like Yo-Yo Ma's Cello—as an Instrument for Human Intelligence" *The Huffington Post*, September, 2015, https://www.huffingtonpost.com/nicholas-agar/robots-human-intelligence_b_8017704.html.

2. Daniel Wegner and Kurt Gray, *The Mind Club: Who Thinks, What Feels, and Why It Matters* (New York: Viking, 2016).

3. John Searle, "Minds, Brains, and Programs," *Behavioral and Brain Sciences* 3 (1980): 417–424.

4. David Chalmers, *The Conscious Mind: In Search of a Fundamental Theory* (Oxford: Oxford University Press, 1996) uses the logical possibility of zombies bereft of phenomenal consciousness to argue that consciousness is a partially nonphysical process.

5. Masahiro Mori, "The Uncanny Valley: The Original Essay," *IEEE Spectrum*, June 12, 2012, https://spectrum.ieee.org/automaton/robotics/humanoids/the-uncanny-valley.

6. Paul Clinton, "Review: 'Polar Express' a Creepy Ride; Technology Brilliant, but Where's the Heart and Soul?" *CNN Entertainment*, November 10, 2004, http://edition.cnn.com/2004/SHOWBIZ/Movies/11/10/review.polar.express.

7. John Cacioppo and William Patrick, *Loneliness: Human Nature and the Need for Social Connection* (New York: W. W. Norton, 2008; Kindle).

8. Ibid., loc. 4119–4120.

9. Paul Seabright, *The Company of Strangers: A Natural History of Economic Life: Revised Edition* (Princeton: Princeton University Press, 2010).

10. Ibid., 4.

11. Ibid., 12.

12. Angela Bahns, Kate Pickett, and Christian Crandall, "Big Schools, Small Schools and Social Relationships," *Group Processes and Intergroup Relations* 15, no. 1 (2012): 119–131.

13. David Johnson and Roger Johnson, "Effects of Cooperative, Competitive, and Individualistic Learning Experiences on Cross-Ethnic Interaction and Friendships," *Journal of Social Psychology* 118, no. 1 (1982): 47–58.

14. Derek Thompson, "A World without Work," *Atlantic Monthly*, July/August 2015, https://www.theatlantic.com/magazine/archive/2015/07/world-without-work/395294.

15. Andrew McAfee and Erik Brynjolfsson, *Machine, Platform, Crowd: Harnessing Our Digital Future* (New York: W. W. Norton and Company, 2017).

16. Danny Lewis, "Reagan and Gorbachev Agreed to Pause the Cold War in Case of an Alien Invasion," *Smithsonian*, November 25, 2015, https://www.smithsonianmag.com/smart-news/reagan-and-gorbachev-agreed-pause-cold-war-case-alien-invasion-180957402/#TIMHr0J7ae5ZF2XR.99.

Chapter 6: Features of the Social Economy in the Digital Age

1. "What Fairtrade Does," Fairtrade, http://fairtrade.org.nz/en-nz/what-is-fairtrade/what-fairtrade-does.

2. Jeremy Rifkin, *The Zero Marginal Cost Society: The Internet of Things, the Collaborative Commons, and the Eclipse of Capitalism* (New York: Palgrave Macmillan, 2014).

3. See Geoffrey Parker, Marshall van Alstyne, and Sangeet Choudary, *Platform Revolution: How Networked Markets Are Transforming the Economy and How to Make Them Work for You* (New York, W. W. Norton and Co, 2016).

4. Antonio Garcia Martinez, *Chaos Monkeys: Obscene Fortune and Random Failure in Silicon Valley* (New York: Harper Collins, 2016).

5. Milton Friedman, "The Social Responsibility of Business Is to Increase Its Profits," *New York Times Magazine*, September 13, 1970, 32–33, 122–124.

6. Lalithaa Krishnan, "The Mark of Exclusivity," *The Hindu*, March 1, 2012, http://www.thehindu.com/arts/crafts/the-mark-of-exclusivity/article2949576.ece.

7. Ed Rensi, "Thanks to 'Fight for $15' Minimum Wage, McDonald's Unveils Job-Replacing Self-Service Kiosks Nationwide," *Forbes*, November 29, 2016, https://www.forbes.com/sites/realspin/2016/11/29/thanks-to-fight-for-15-minimum-wage-mcdonalds-unveils-job-replacing-self-service-kiosks-nationwide.

8. Douglas Rushkoff describes digiphrenia—"the experience of trying to exist in more than one incarnation of yourself at the same time. There's your Twitter profile, there's your Facebook profile, there's your email inbox. And all of these sort of multiple instances of you are operating simultaneously and in parallel." People afflicted with digiphrenia can be located in many different places at once. Rushkoff observes that it's "not a really comfortable position for most human beings." Douglas Rushkoff, "In a World That's Always On, We Are Trapped in the 'Present,'" *All Things Considered*, NPR, https://www.npr.org/2013/03/25/175056313/in-a-world-thats-always-on-we-are-trapped-in-the-present.

9. Sherry Turkle, *Reclaiming Conversation: The Power of Talk in the Digital Age* (New York: Penguin Books, 2015; Kindle), loc. 47.

10. Ibid., loc. 4.

11. Leslie Hook, "Amazon to Launch Checkout-free Offline Grocery Store," *Financial Times*, December 6, 2016, https://www.ft.com/content/88399c00-bb03-11e6-8b45-b8b81dd5d080.

12. "Pretty Woman," Wikiquote, https://en.wikiquote.org/wiki/Pretty_Woman.

13. Brad Stone, *The Upstarts: How Uber, Airbnb, and the Killer Companies of the New Silicon Valley Are Changing the World* (Boston: Little, Brown and Co.; 2017; Kindle), loc. 3656.

14. Quoted in Stone, *The Upstarts*, Kindle loc. 3663.

15. Fitz Tepper, "Uber Has Completed 2 Billion Rides," *Techcrunch*, July 18, 2016, https://techcrunch.com/2016/07/18/uber-has-completed-2-billion-rides.

16. Stone, *The Upstarts*.

17. Quoted in Stone, *The Upstarts*, Kindle loc. 4974.

18. Quoted in Stone, *The Upstarts*, Kindle loc. 4981.

19. Bethany McLean and Peter Elkind, *The Smartest Guys in the Room: The Amazing Rise and Scandalous Fall of Enron* (New York: Portfolio Trade, 2004).

20. "What Are Airbnb Service Fees?," Airbnb, https://www.airbnb.co.nz/help/article/104/what-are-guest-service-fees.

21. Gemma Lavender, "Why Send People into Space When a Robotic Spacecraft Costs Less?," Space Answers, May 27, 2015, https://www.spaceanswers.com/space-exploration/why-send-people-into-space-when-a-robotic-spacecraft-costs-less.

22. "Apollo 11 Moon Landing: Ten Facts about Armstrong, Aldrin, and Collins' Mission," *Telegraph*, July 18, 2009, http://www.telegraph.co.uk/news/science/space/5852237/Apollo-11-Moon-landing-ten-facts-about-Armstrong-Aldrin-and-Collins-mission.html.

23. Michael Ryan, "A Ride in Space," *People*, June 20, 1983, http://people.com/archive/cover-story-a-ride-in-space-vol-19-no-24.

Chapter 7: A Tempered Optimism about the Digital Age

1. See Evgeny Morozov, "Data Populists Must Seize Our Information—for the Benefit of Us All," *Guardian*, December 4, 2016, https://www.theguardian.com/commentisfree/2016/dec/04/data-populists-must-seize-information-for-benefit-of-all-evgeny-morozov.

2. Jeremy Rifkin, *The Zero Marginal Cost Society: The Internet of Things, the Collaborative Commons, and the Eclipse of Capitalism* (New York: Palgrave Macmillan, 2014).

3. Paul Romer, "Conditional Optimism about Progress and Climate," https://paul romer.net/conditional-optimism-about-progress-and-climate.

4. Rifkin, *The Zero Marginal Cost Society*.

5. Ibid., 18.

6. Alvin Toffler, *The Third Wave: The Classic Study of Tomorrow* (New York: Bantam, 1980).

7. Tim Mullaney, "Jobs Fight: Haves vs. the Have-Nots," *USA Today*, September 16, 2012, http://usatoday30.usatoday.com/money/business/story/2012/09/16/jobs -fight-haves-vs-the-have-nots/57778406/1.

8. Andrew Keen, *The Internet Is Not the Answer* (London: Atlantic Books, Kindle), loc. 180.

9. For an account of the impressive powers of production the United States committed to the production of weapons, see A. J. Baime, *The Arsenal of Democracy: FDR, Detroit, and an Epic Quest to Arm an America at War* (New York: Mariner Books, 2015).

10. Chris Anderson, *Makers: The New Industrial Revolution* (New York: Crown Business, 2012).

11. Jathan Sadowski, "Why Silicon Valley Is Embracing Universal Basic Income," *Guardian*, June 22, 2016, https://www.theguardian.com/technology/2016/jun/22 /silicon-valley-universal-basic-income-y-combinator.

12. For book-length philosophical defenses, see Philippe Van Parijs, *Real Freedom for All: What (If Anything) Can Justify Capitalism?* (Oxford: Oxford University Press, 1997), and more recently, Mark Walker, *Free Money for All: A Basic Income Guarantee Solution for the Twenty-First Century* (Basingstoke, UK: Palgrave Macmillan, 2016).

13. Martin Ford, *Rise of the Robots: Technology and the Threat of a Jobless Future* (New York: Basic Books, 2015; Kindle), loc. 4231.

14. See Scott Santens, "A Future Without Jobs Does Not Equal a Future Without Work," *Huffington Post*, October 7, 2016, https://www.huffingtonpost.com/scott -santens/a-future-without-jobs-doe_b_8254836.html.

15. Van Parijs, *Real Freedom for All*.

16. Walker, *Free Money for All*.

17. Byron Reese, *Infinite Progress: How the Internet and Technology Will End Ignorance, Disease, Poverty, Hunger, and War* (Austin, TX: Greenleaf Book Group Press, 2013), 102.

Chapter 8: Machine Breaking for the Digital Age

1. Most notably, E. P. Thompson, *The Making of the English Working Class* (London: Penguin, 1980; Kindle).

2. Mark Zuckerberg, "Building a Global Community," Facebook, https://www.facebook.com/notes/mark-zuckerberg/building-global-community/10103508221158471/?pnref=story.

3. Brian Solomon, "Airbnb Raising More Cash at $30 Billion Valuation," *Forbes* September 22, 2016, https://www.forbes.com/sites/briansolomon/2016/09/22/airbnb-fundraising-850-million-30-billion-valuation.

4. Kevin Montgomery, "Airbnb Thinks It Should Win the Nobel Peace Prize," *Valleywag,* November 21, 2014, http://valleywag.gawker.com/airbnb-thinks-it-should-win-the-nobel-peace-prize-1661900628.

5. Andrew Sorkin, "The Mystery of Steve Jobs's Public Giving," *New York Times,* August 29, 2011, https://dealbook.nytimes.com/2011/08/29/the-mystery-of-steve-jobss-public-giving.

6. Jason Snell, "Steve Jobs: Making a Dent in the Universe," *Macworld,* October 6, 2011, https://www.macworld.com/article/1162827/macs/steve-jobs-making-a-dent-in-the-universe.html.

7. Observer Editorial, "The Observer View on Mark Zuckerberg" *Observer,* February 19, 2017, https://www.theguardian.com/commentisfree/2017/feb/19/the-observer-view-on-mark-zuckerberg.

8. Brad Stone, *The Everything Store: Jeff Bezos and the Age of Amazon* (New York, Little, Brown, and Company, 2013).

9. See Evgeny Morozov, "Data Populists Must Seize Our Information," *Guardian,* December 4, 2016, https://www.theguardian.com/commentisfree/2016/dec/04/data-populists-must-seize-information-for-benefit-of-all-evgeny-morozov.

10. Amanda Schaffer, "Tech's Enduring Great-Man Myth," *MIT Technology Review,* August 4, 2015, https://www.technologyreview.com/s/539861/techs-enduring-great-man-myth.

11. Mary Aiken, *The Cyber Effect: A Pioneering Cyberpsychologist Explains How Human Behavior Changes Online* (New York, Random House, 2014; Kindle 2014), loc. 2308.

12. Milton Friedman, "The Social Responsibility of Business Is to Increase Its Profits," *New York Times Magazine,* September 13, 1970, 32–33, 122–124.

13. Theodore Schleifer, "Uber's Latest Valuation: $72 Billion," *Recode,* February 9, 2018, https://www.recode.net/2018/2/9/16996834/uber-latest-valuation-72-billion-waymo-lawsuit-settlement.

14. Robert Reich, "The Share-the-Scraps Economy," *Huffington Post*, February 2, 2015, https://www.huffingtonpost.com/robert-reich/the-sharethescraps-econom_b_6597992.html.

15. This seems to be one union response to Uber. See "We Are Uber, Lyft, Juno, Via Workers United for a Fair Industry," Independent Drivers Guild, https://drivingguild.org.

16. Maya Kosoff, "Everything You Need to Know about *The Fountainhead*, a Book That Inspires Uber's Billionaire CEO Travis Kalanick," *Business Insider*, June 1, 2015, http://www.businessinsider.com.au/how-ayn-rand-inspired-uber-ceo-travis-kalanick-2015-6.

17. Quoted in Brad Stone, *The Upstarts: How Uber, Airbnb, and the Killer Companies of the New Silicon Valley Are Changing the World* (Boston: Little, Brown and Co., 2017; Kindle), loc. 3663.

Chapter 9: Making a Very Human Digital Age

1. John Cacioppo and William Patrick, *Loneliness: Human Nature and the Need for Social Connection* (New York: W. W. Norton, 2008).

2. Paul Seabright, *The Company of Strangers: A Natural History of Economic Life*, Revised Edition (Princeton, NJ: Princeton University Press, 2010), 4.

3. Julie Miller, "Matt Damon May Have Made $1 Million per Line in *Jason Bourne*," *Vanity Fair*, July 18, 2016, https://www.vanityfair.com/hollywood/2016/07/matt-damon-jason-bourne-pay.

Index

Aiken, Mary, 182
Airbnb, 145–149, 176, 186
Amazon, 44, 57, 61, 64, 91–92, 137, 143–144, 175, 177, 178
Anderson, Chris, 164
Andreessen, Marc, 159, 165
Anthropocene Epoch, 196, 197
Ashley Madison (online dating service), 54–55
Autor, David, 36–38, 86–92, 99–100, 193–194

Bahns, Angela, 120
Bacon, Francis, 1, 196
Belief in human exceptionalism (bias in assessing progress in artificial intelligence), 6–9, 81, 94–96, 150–151
Bezos, Jeff, 64, 74, 177, 178
Black Mirror (TV series), 111–112, 128
Bostrom, Nick, 5
Brand, Stewart, 18, 68–70
Brynjolfsson, Erik, 30–31, 87, 97, 124–125

Cacioppo, John, 10–11, 118, 194
CanceRx (machine learning approach to cancer), 39–41, 44, 47, 58, 61, 97–98, 101
Chatbot (computer program that converses with humans), 52–56, 124
Chesky, Brian, 147
Childe, V. Gordon, 23–26
Chinese Room Argument (argument due to John Searle), 49–50, 114–115
Climate change, 5, 6–7, 15, 158, 177–178, 185
Collaborative commons (view of the digital economy due to Jeremy Rifkin), 69–70, 155–156, 157–158, 163–165, 173
Crandall, Christian, 120
Csikszentmihalyi, Mihaly, 83–84, 122

Doctorow, Cory, 70
Domingos, Pedro, 38–41, 44, 47–51, 93
Dorn, David, 90–92, 100

ELIZA (1960s natural language processing computer program), 54–55

Facebook, 11, 12, 18, 25, 29, 44, 61–68, 71–79, 101, 135, 142, 146, 157, 159, 164, 175, 176–179, 183, 197
Ford, Martin, 166–167, 170, 172
Friedman, Milton, 139–140, 176–177, 182–183, 187
Friedman, Thomas, 31

Garcia Martinez, Antonio ("a billion times anything is still a big number"), 73, 135
Gigerenzer, Gerd, 36–37

Google, 18, 31, 32, 37, 44, 61, 63–65, 67, 71–72, 73–79, 139, 146, 159, 164, 175, 178–179, 197
Gordon, Robert, 16–17, 26–42, 178
Gray, Kurt, 10

HAL 9000 (malfunctioning homicidal computer in Stanley Kubrick's movie *2001: A Space Odyssey*), 46–47
Hammerbacher, Jeff, 32–33
Hyperactive Agency Detector (a hypothesized evolved bias toward detecting agency), 54–57, 59, 124, 133

Insurance mindset (as a collective response to future uncertainty), 102–107, 191, 191–193

Jefferson, Thomas, 69

Kalanick, Travis, 146, 187–189
Kasparov, Garry, 50–51, 58, 152, 204
Keen, Andrew, 159, 165
Kelly, Kevin, 18, 70, 72, 84–86
Keynes, John Maynard, 85
Krishnan, Lalithaa, 141
Kurman, Melba, 33–34

Lanier, Jaron, 18, 67, 73–79, 183
LeFevre, Judith, 83–84, 122
Levesque, Hector, 53, 56, 204
Life-dinner principle, 76–78
Lipson, Hod, 33–34
Loebner Prize, 52–56, 124, 204–205
Long view of technological change (a view of technological change that prioritizes long-term trends), 2–3, 6, 7, 17–18, 23–26, 42, 43–44, 68, 74, 85, 99, 112, 148, 171, 189

McAfee, Andrew, 30–31, 87, 97, 124–125
Machine learning, 4–5, 26, 33–41, 47–49, 51, 65, 81–82, 89, 92–93, 110, 128, 178, 181, 193
Machine learning, 4, 26, 34, 36–41, 47–49, 51, 65, 81–82, 89–93, 110, 128, 178, 181, 193, 204
Merely counterfactual bias (bias against currently non–existent beings), 126–129
Micropayment (small sum of money paid for the use of online content), 18, 73–79, 183
Mind club, 10, 19, 44–57, 106, 109–112, 124–126
Mind work, 4, 9, 17–18, 31, 43–53, 59–60, 61, 81, 89, 90, 94, 100, 106, 132, 165–166, 168–169, 172, 181, 191–192
Mithen, Steve, 24
Mori, Masahiro, 117–118
Morozov, Evgeny, 212, 214
Mukherjee, Siddhartha, 39–40

Nasmyth, James, 82

O-ring production function (as part of an argument for indispensable human contributions), 86–88, 92, 99, 106

Patrick, William, 10–11, 118, 194
Phenomenal consciousness, 16, 112–119
Pickett, Kate, 120
Present bias about digital technologies (bias about progress in artificial intelligence), 6–9, 81, 94–96, 150–151
Pro-human bias (bias in favor of interacting with other humans), 125–129
Protean technology (the steam engine and networked computer as examples), 18, 41, 82, 88–92, 102, 107, 109
Putnam, Robert, 11

Ride, Sally, 152
Rifkin, Jeremy, 18, 69–70, 135, 155–159, 163–165, 174
Romer, Paul, 156–158
Rosen, William, 201
Rushkoff, Douglas, 211

Sadowski, Jathan, 165
Schaffer, Amanda, 178
Schneier, Bruce, 78–79
Seabright, Paul, 119–120, 194
Searle, John, 49–50, 114–115
Social economy, 10–14, 98, 108–110, 123–125, 129, 131 153, 155, 159–160, 165, 173, 175–176, 181, 193–194, 196–198
Stallman, Richard Matthew, 69
Star Trek, 8, 45, 125–127
Stepford Wives, The, 112
Stone, Brad, 145–147

Technological package, 21, 24–26, 196–198
Technological unemployment, 85
Tesla, 33, 34, 37, 93
Thompson, Derek, 122
TINA (There is no alternative), 179–181
Toffler, Alvin, 157
Turing, Alan, 1, 17, 43–46, 49, 51–60, 61, 110, 171
Turkle, Sherry, 142–143
23andMe, 18, 38, 66, 68, 72, 75

Uber, 137, 145–149, 185–189
Uncanny valley (sense of uneasiness about things that are almost human), 117–118
Universal Basic Income (UBI), 14, 20, 121, 156, 163, 165–173, 194–196

Van Parijs, Philippe, 170–171

Walker, Mark, 171
Wegner, Daniel, 10
Work norm (the idea that it is normal, but not universal, for humans to grow up and find work), 14, 82–84, 92, 99–107, 156, 191–196
Worker platforms (proposed response by workers to platform businesses like Uber), 185–189
Wozniak, Steve, 5–6, 165

Zuckerberg, Mark, 41, 65, 68, 73, 101, 159, 176–178